ECOLOGY AND EQUITY

Ecology and Equity presents a provocative interpretation of the environment debate in a large, diverse and vitally important Third World country. Its focus is not so much on the extent of environmental degradation in India as on its manifold human consequences, using an original theoretical framework for making sense of what is, from an ecological point of view, undoubtedly the most complex society in the world.

The authors divide this society into three categories: omnivores, ecosystem people and ecological refugees. The processes of environmental degradation and social conflict are analysed in terms of the inequities in access to natural resources by these competing classes. These conflicts have fuelled a vibrant environmental movement, of groups varied in ideological affiliation and strategies of action.

The book then turns from analysis to prescription, arguing for an environment-friendly agenda for development. This agenda – termed 'Conservative-Liberal-Socialism' – creatively combines elements from different political traditions. Its principles are developed for several key sectors, such as forestry, information and population.

Ecology and Equity provides the first analytically sophisticated and empirically grounded interpretation of environmental conflict in India. The book innovatively combines political economy with ecology, emphasizing a forward-looking agenda for environmental reform in the Third World.

Madhav Gadgil is a Professor in the Centre for Ecological Sciences at the Indian Institute of Science; **Ramachandra Guha** is an independent writer. They previously collaborated on *This Fissured Land: An Ecological History of India* (1992).

ECOLOGY AND EQUITY

The use and abuse of nature in contemporary India

Madhav Gadgil and
Ramachandra Guha

London and New York

First published 1995
by Routledge
11 New Fetter Lane, London EC4P 4EE

Simultaneously published in the USA and Canada
by Routledge
29 West 35th Street, New York, NY 10001

Typeset in Garamond by
Ponting–Green Publishing Services, Chesham, Bucks
Printed and bound in Great Britain by
Biddles Ltd, Guildford and King's Lynn

British Library Cataloguing in Publication Data
A catalogue record for this book is available from the
British Library

Library of Congress Cataloguing in Publication Data
A catalogue record for this book has been requested

ISBN 0–415–12523–5
0–415–12524–3 (pbk)

To the memory of

Jotiba Phule
J.C. Kumarappa
Mira Behn
Salim Ali
Abdul Nazir Saab

(The tragopan to the Bigshot)

'You say the town is short of water
Yet at the wedding of your daughter
The whole municipal supply
Was poured upon your lawns. Well, why?
And why is it that Minister's Hill
And Babu's Barrow drink their fill
Through every season, dry or wet,
When all the common people get
Is water on alternate days?
At least, that's what my data says,
And every figure has been checked.
So, Bigshot, wouldn't you expect
A radical redistribution
Would help provide a just solution?'

Vikram Seth,
The Elephant and the Tragopan

Right
I will not stop cutting down trees
Though there is life in them
I will not stop plucking out leaves,
Though they will make nature beautiful
I will not stop hacking off branches,
Though they are the arms of a tree
Because –
I need a hut

Cherabandaraju
(translated from Telugu by C.V. Subbarao)

CONTENTS

List of plates viii
Acknowledgements x
UNRISD xi

INTRODUCTION 1

Part I The India that is

1 CORNERING THE BENEFITS 9
2 PASSING ON THE COSTS 34
3 A CAULDRON OF CONFLICTS 61
4 IDEOLOGIES OF ENVIRONMENTALISM 98

Part II The India that might be

5 CONSERVATIVE-LIBERAL-SOCIALISM 115
6 KNOWLEDGE OF THE PEOPLE, BY THE PEOPLE, FOR THE PEOPLE 133
7 WHAT ARE FORESTS FOR? 148
8 IS THERE SAFETY IN NUMBERS? 176
9 RESOURCES OF HOPE 184

Glossary of words in Indian languages 192
Glossary of words referring to Indian communities 195
Bibliography 196
Index 204

LIST OF PLATES

1 Ecological refugees in urban India struggling for water 14
2 India's elite now reach out to the remotest parts of the country 15
3 Mechanized fisheries and chilling and canning plants support a major export industry 16
4 River valley projects have triggered deforestation 21
5 Fly ash blankets the land near coal-fired power plants 22
6 Railway line overhanging the landslide it has caused 30
7 Untreated sewage from Bangalore poses a serious health hazard 31
8 Shantytowns of ecological refugees 32
9 A fuelwood depot in Karnataka 40
10 Before extraction of the ore, natural grassy banks and evergreen forest graced the site of the Kudremukh mines 44
11 The Kudremukh iron ore mines have laid to waste one of the most picturesque parts of the Western Ghats 44
12 Unregulated removal of sand from river beds has led to disruption of water regimes 57
13 Medha Patkar leading a *dharna* by Narmada Bachao Andolan supporters and tribal people 62
14 There is inadequate control of industrial pollution 79
15 Rapid mechanization of fisheries has led to overexploitation of fish stocks 82
16 Mechanized fisheries have become an important component of the economy of the Andaman and Nicobar Islands 82
17 The site of an abandoned iron mine which once supported primeval rain forest 90
18 Unregulated quarrying of granite is a significant cause of deforestation and disruption of watercourses 90
19 Women carrying fuelwood out of the Ranebennur wildlife sanctuary in Karnataka 93
20 A procession through Harsud, a town that was about to be submerged under the rising waters of the Narmada 106

21 Tropical rain forests still cover extensive areas of the Western Ghats, a hot spot for biodiversity 150
22 The pristine rain forest of the Andaman and Nicobar Islands – a treasure house of biodiversity 150
23 Forest-based industry is now increasingly dependent on the as yet little-exploited forests of the Andaman and Nicobar Islands 152
24 Hills in dry parts of peninsular India, laid totally bare by charcoal production 154
25 Earthen mound inside which an unusual bird, the Nicobar megapode, lays its eggs 156
26 One of the many species of trees endemic to the Andaman and Nicobar Islands 157
27 Rafts of timber floating down streams of the Andaman and Nicobar Islands 163
28 India's forest-based industry is now tapping the rain forests of the Andaman and Nicobar Islands 164

ACKNOWLEDGEMENTS

This book uses a variety of sources: aside from published materials, many of the illustrative examples of resource use and abuse come from twenty years of field experience on the part of one of the writers (Gadgil), working with ecosystem people, voluntary agencies, people's science movements and different levels of government. We have also benefited from conversations over the years with friends, too numerous to list individually, working in the spheres of academia, activism and administration.

Helpful comments on an earlier draft of the book came from Anil Gore, S. Bharath, Jayakumar Anagol, Keshav Desiraju, R. Sudarshan, Jessica Vivien, Michael Redclift, Jean Dreze, Dharam Ghai and Siddhartha Gadgil. Aside from typing drafts, Vijayageetha Gadagkar and Prema Iyer helped in many other ways in putting the book together; their assistance has been critical. The photographs were kindly supplied by the magazine *Frontline* (Plates 13 and 20) and Father Cecil Saldanha (all others). Finally, we would like to thank the Ministry of Environment and Forests, Government of India, for institutional support; and UNRISD and its Director, Dharam Ghai, for their patience and encouragement.

UNRISD

The United Nations Research Institute for Social Development (UNRISD) is an autonomous agency that engages in multi-disciplinary research on the social dimensions of contemporary problems affecting development. Its work is guided by the conviction that, for effective development policies to be formulated, an understanding of the social and political context is crucial. The Institute attempts to provide governments, development agencies, grassroots organizations and scholars with a better understanding of how development policies and processes of economic, social and environmental change affect different social groups. Working through an extensive network of national research centres, UNRISD aims to promote original research and strengthen research capacity in developing countries.

Current research themes include Crisis, Adjustment and Social Change; Socioeconomic and Political Consequences of the International Trade in Illicit Drugs; Environment, Sustainable Development and Social Change; Ethnic Conflict and Development; Integrating Gender into Development Policy; Participation and Changes in Property Relations in Communist and Post-Communist Societies; Refugees, Returnees and Local Society; and Political Violence and Social Movements. UNRISD research projects focused on the 1995 World Summit for Social Development include Rethinking Social Development in the 1990s; Economic Restructuring and New Social Policies; Ethnic Diversity and Public Policies; and The Challenge of Rebuilding Wartorn Societies.

A list of the Institute's free and priced publications can be obtained by writing to UNRISD, Reference Centre, Palais des Nations, CH-1211, Geneva 10, Switzerland.

INTRODUCTION

I

Snapshots of our living earth relayed by a satellite above have undoubtedly been among the most dramatic images ever encountered by humans. This remarkable image has become commonplace only over the past decade. Let us for a moment imagine that such an image had been available for that momentous year 1947, when India attained freedom. If we were to contrast that fictional image with one captured for us by a satellite today, striking differences would be readily apparent. The earlier image would have much smaller areas of barren rocks and urban sprawl; fewer large bodies of water, but many smaller lakes freer of weeds at the peak of the monsoon; much more extensive tree cover, but a far lower extent of crop fields retaining their greenery even at the height of the summer.

Satellite pictures, real and imagined, help us understand the radical changes in the Indian landscape over the past four decades. It is a landscape in which the natural world is continually being replaced by a world of artefacts: where trees, shrubs and grasses are giving way to plantations and crop fields, roads and buildings; where rivers are being increasingly impounded with waters diverted through underground tunnels to turn giant turbines or merely being disciplined to flow along paths straight and narrow; where old wetlands are being drained and new ones created in the form of waterlogged fields.

For a fuller appreciation of the *human* consequences of this transformation, however, the modern bird's-eye view of the satellite might be supplemented by a more traditional worm's-eye view of life on the ground. A rapid walking tour through the cities and villages of India can help us understand how changes in the landscape reflect more subtle changes in the social fabric, and the ways in which India's intricate mosaic of human groups is coping with these changes.

A tourist walking through the Indian countryside (at rather more than a worm's pace) would be immediately struck by the chronic shortages of natural resources faced by every segment of Indian society. Fisherfolk are faced with the exhaustion of fish stock, shifting cultivators with the declining availability of forest land. Mat weavers are running short of reeds and peasants are short of dung with which to manure their fields. Millions among the urban poor are shelterless and without adequate water supply. Irrigated farmlands are turning saline and whole coconut orchards are dying of disease. Paper mills are starved of their favourite raw material, bamboo, and textile mills are plagued by power cuts. City roads are clogged with traffic and city air is full of noxious fumes. The ever-growing numbers of Indians, their exploding appetite for consumption and their wasteful patterns of resource use have together conspired to ensure that all segments of society are in the midst of one resource crunch or another.

If the bird's-eye view revealed a picture of considerable ecological change, the worm's-eye view converts that image into one of a serious ecological crisis. This crisis is being translated into increasing social conflict, as different groups exercise competing claims on a dwindling resource base. India today is a veritable cauldron of social conflicts, many of which pertain directly to the control and use of natural resources. These conflicts are played out at different levels and with varying intensities. Within numerous scattered small villages, rich farmers and landless labourers fight for access to common grazing ground, while in city slums desperately poor households quarrel over the trickle of water that reaches them from a sole municipal tap. Such localized conflicts usually go unreported, and far better known are resource conflicts that involve large numbers of people and occur over extensive areas. These include, for example, the massive displacement of villages by a chain of dams being built on the Narmada river in central India, or the bitter fight between the politicians of the states of Karnataka and Tamil Nadu over the waters of the river Kaveri.

These conflicts, localized within villages or spread across large regions, provide the backdrop to the vibrant environmental movement in India. This is a movement that has grown rapidly in the past two decades. Indeed, it is the protests of environmentalists, rather than the concerns of the state or of the intelligentsia, that have generated a wide public awareness of the extent of environmental degradation in India, and (what is more important still) of its human consequences.

The numerous local groups comprising the environmental movement have been concerned, above all, with stopping economic activities that destroy the environment and impoverish local communities – be they large dams on the Narmada or magnesite mines in the inner Himalaya. By its very nature this has been a defensive movement, at times little more than a holding operation. In the circumstances it is hardly surprising that the environmental movement in India has not given sufficient thought to the larger processes that are contributing both to ecological deterioration and to social strife. Here some environmentalists have focused too narrowly on individual actors – for example, on forest managers in the case of the Chipko movement against commercial forestry, or on the World Bank as in the case of the Narmada agitation – while others have been content with identifying impersonal, abstract forces such as 'capitalism', 'materialism' or even 'modern Western patriarchal science' as being ultimately responsible for our present predicament. Universally lacking is a proper social-scientific analysis that might locate these individual actors in a wider context, or which would give flesh and bone to broad concepts such as 'capitalism' or 'science'.

II

The first part of *Ecology and Equity* presents an original theoretical framework for making sense of what is, from an ecological point of view, undoubtedly the most complex society in the world. This is a society that contains within its ranks Stone-Age hunter-gatherers of the Andamans and white-collar *babus* of Delhi, nomadic shepherds of Himachal Pradesh and pavement dwellers of Calcutta, artisanal fisherfolk of Tamil Nadu and purse seine operators of Goa, shifting cultivators of Mizoram and sugar barons of Maharashtra, textile mill owners of Coimbatore and software exporters of Bangalore, fuelwood headloaders of Kumaon and engineers drilling the Bombay High for offshore oil. These varied constituents of Indian society differ greatly among themselves in their access to the resources of the earth. While bureaucrats in Delhi daily watch American soap operas on their Japanese television sets, the vast majority of women in villages of Bihar and Rajasthan can neither read, nor even listen to a cheap transistor radio. While citizens of Bombay have drinking water brought to their taps from rivers dammed tens of kilometres away, women in villages of Saurashtra must trudge long distances to bring home a pitcher of water. While obesity clinics sprout up in Madras and Bangalore, fully one-third of the Indian people cannot afford to buy enough food to keep their body and soul together.

The relentless transformation of the natural world into the world of artefacts, brought out so vividly for us by the satellite, has most asymmetric implications for these different constituents of Indian society. For the many who earn barely enough to fill their bellies, there is little left over to acquire the new goods on the market, be they soaps or blenders, mopeds or TV sets, apples flown in from the Himalaya or flats in high-rise buildings. The bulk of the poor, or even the not-so-affluent, must scratch the earth and hope for rains in order to grow their own food, must gather wood or dung to cook it, must build their own huts with bamboo or sticks of sorghum dabbed with mud and must try to keep out mosquitoes by engulfing them with smoke from the cooking hearth. Such people depend on the natural environments of their own locality to meet most of their material needs. Perhaps four-fifths of India's rural people, over half of the total population, belong to this category, which, following Raymond Dasmann (1988), we call *ecosystem people*.

As the natural world recedes, so shrink the capacities of local ecosystems to support these people. Dams and mines, for instance, have physically displaced millions of peasants and tribals in independent India. Others have fled as forests and, with them, springs have vanished. These people constitute the *ecological refugees* who live on the margins of islands of prosperity, as sugarcane harvesters in western Maharashtra or farm labourers in Punjab, as hawkers and domestic servants of Patna or Hyderabad. As many as one-third of the Indian population probably live today such a life as displacees, with little that they can freely pick up from the natural world, but not much money to buy the commodities that the shops are brimming with either.

The remaining one-sixth of India's population are the real beneficiaries of economic development, which might be defined for the present purpose as the growth of the artificial at the cost of the natural. These beneficiaries are bigger landowners with access to irrigation, these are modern entrepreneurs in pockets of industrialization and the workers in the organized sector, these are the urban professionals – lawyers, doctors, investment bankers – rapidly gaining in wealth and prestige, and these are the ever-growing numbers of employees in government, semi-government and government-aided organizations. They have the purchasing power to buy cars and fly in airplanes, to dress in polyester clothes and feast on the fish, flesh and fruit brought to them from the four corners of the land. Not only do they have the money to pay for these commodities, but they have the clout to use the power of the state to ensure that the goodies come to them cheap, if not altogether free. As prosperous farmers they pay next to nothing for the electricity that runs their pump sets, as city dwellers they pay little for the water brought to them from hundreds of kilometres away. The news they read is printed on paper subsidized by the low rate at which bamboo is supplied to the mills, and the state builds and maintains at its own expense the highways on which ply the lorries that bring them all manners of commodities from great distances. Like their Western counterparts, whom Raymond Dasmann calls 'biosphere people', they enjoy the produce of the entire biosphere, in contrast to the ecosystem people, who have a very limited resource catchment. Devouring everything produced all over the earth, they might equally be termed *omnivores*.

Omnivores, *ecosystem people* and *ecological refugees*: three broad categories, to which we might assign all of India's huge population. These three classes might be distinguished by the size of their respective resource catchments, or by their relative ability to transform nature into artefact. They might be distinguished too by their widely varying powers to influence state policy, or by the degree of control they exercise over their own lives. These are thus categories at once ecological and sociological, and we use them in this book to provide a fresh interpretative analysis of the development experience in India, and of the social conflicts that have come in its wake.

Like all attempts to classify and interpret complex phenomena, this one too would run into the inevitable boundary problems. Is a village schoolteacher, with an assured salary from the state exchequer and perhaps some land, too, an ecosystem person, or is he a rural omnivore? The owner of a small garage in the city lacks the social power to qualify as an authentic omnivore, but can hardly be called an ecological refugee either. These examples could be multiplied; but while recognizing this difficulty, we are nevertheless convinced that a majority of the Indian population is covered by the three categories identified by us. More crucially, from a socio-ecological point of view we believe our categories to be a great improvement on the more conventional ones of class or interest group. This is not to say that the frameworks of class and interest group cannot be fruitfully applied in the study of other societies or other historical contexts. But as we hope to show in this book, our alternative, threefold classification provides a fuller and more convincing interpretation of political, economic and environmental change in contemporary India.

III

In the second part of *Ecology and Equity* we turn from analysis to reconstruction, arguing for a new environment-friendly agenda for development that we believe to be in the interest of a vast majority of the Indian people. In their own, undoubtedly sincere, opposition to large projects, environmental groups have not thus far spelt out any concrete alternatives to present processes of destruction and deprivation. This might only be consistent with the defensive, almost siege-like position they find themselves in, but environmentalists have not always helped their cause by appearing to Just Say No to everything – be it eucalyptus, large dams or modern science. It has thus been easy for their opponents to dub them as anti-development, as backward-looking, retrograde rabble-rousers.

Despite its vitality and rapid rise to prominence, then, the environmental movement has been unable to contribute creatively to major debates on development policy in contemporary India. These debates have been conducted in the main between the proponents of economic liberalization, who favour a market economy and an outward-looking India fully integrated with the global marketplace; and those who wish to preserve the *status quo ante*, where the state has occupied the commanding heights of the economy and where a largely self-reliant India stands gloriously isolated, economically speaking, from the rest of the world. Yet this is a debate that might be greatly enriched, and given more meaning and relevance to the Indian context, by providing it with an ecological perspective. This does not mean simply balancing the sometimes conflicting objectives of economic growth and environmental protection, but rather, integrating into every step of policy formulation and execution the insights gleaned from an ecological interpretation of the Indian development experience. That, precisely, is what is attempted in the later chapters of this book.

We firmly believe that prudent, sustainable use of India's environmental resources is in the interests of a vast majority of India's population. It is therefore entirely possible to construct a development agenda that is at once in the interests of a majority of people and of the country's environment. Since most people appear to be primarily engaged in a pursuit of self-interest, such a programme ought to attract broad-based support. The snag, of course, is that the small minority of people who stand to lose from such a development process are the very people who are entrenched in power today. Therefore the agenda we advocate is bound to run into powerful opposition, at home as well as abroad. But we have faith that in the long run Indian society can move with vigour in the direction of the alternative future that we sketch in the second half of the book.

In this manner, *Ecology and Equity* hopes to provide the Indian environmental movement with a fuller analysis of the processes it has been fighting against; and, what is perhaps more important still, with a vision of what it should be fighting for. A truer picture, then, of the India that is, and a dream of the India that might be.

Part I

THE INDIA THAT IS

1

CORNERING THE BENEFITS

THE COLONIAL LEGACY

India is a natural geographical entity bounded by the great ranges of the Himalaya, the Thar desert and the Indian Ocean. Over the centuries, people have flowed in and out of it through passages to the northwest, the northeast and across the seas; but cultural exchanges have been especially intense within the boundaries of the subcontinent. While India has thus been a cultural entity for over two millennia, it was politically constituted as a state under the British colonial regime only about two hundred years ago.

The British who conquered and unified India were at that time the world's premier omnivores, drawing resources of the entire biosphere to their tiny island kingdom. The men presiding over the British Empire perched on chairs of Burma teak at tables of African mahogany, consuming Australian beef washed down with French and Italian wines. Their women were decked in Canadian furs and clothes of Egyptian cotton, dyed with Indian indigo, glittering with diamonds from South Africa and gold from Peru.

These levels of resource consumption among the British elite were attained by draining their many colonies, including of course India, of their natural resources. In order to accomplish this the pattern of land use within India had been so organized as to maximize the revenue it yielded to the British crown, and the commodities it could produce to feed the British economy. Since village communities could not conveniently be held responsible to pay taxes, land either became private property or was taken over by the crown. The privately held lands were primarily cultivated lands that were taxed heavily. In much of north and east India the ownership was handed over to feudal landlords, with the peasantry reduced to the status of much-exploited tenants and sharecroppers. In parts of south and west India cultivators were assigned lands, but, unable to pay the high taxes, quickly became chronically indebted, losing their lands to moneylenders. The peasants were forced to cultivate cash crops such as cotton, jute and indigo to feed the expanding British textile industry; but whatever the crops cultivated, they often could not make ends meet. While more and more land was brought under cultivation, the productivity of agriculture remained utterly stagnant under the colonial regime. There were several disastrous famines under British rule with millions of deaths. With population beginning to grow after the First World War, per capita food production in India was at its lowest ever at the conclusion of the Second World War (see Guha 1992).

Peasants in India cultivate tropical soils, most of which are poor in nutrients. They compensate for this by maintaining a herd of cattle that convert the vegetation on the surrounding non-cultivated lands into manure for the crops. Peasant families must also cook their meals with fuelwood, and construct their huts and weave their baskets from the small timber and bamboo gathered from these lands. Such lands, as well as irrigation ponds, used to be managed for the most part by village communities. This involved restraints on overuse and contributions to maintenance, such as the periodic desilting of ponds by communal labour. However, the British had scant sympathy for community-based management systems. Where they contributed to higher levels of agricultural taxation as with irrigation ponds, they were permitted to continue. Otherwise, as with wood lots and grazing lands, the state took them over, declaring community control illegitimate. These lands were then either dedicated to producing timber on reserved forest lands or were constituted into open-access lands that suffered overuse and degradation. While in theory the reserved forest lands were supposed to be managed in a sustainable fashion, their takeover by the state merely amounted to confiscation, not conservation, as a dissenting colonial official eloquently put it in the 1870s. The tremendously diverse tree growth of India's forests was liquidated to build British ships and lay extensive railway lines. Mixed forests were replaced by single-species stands of a handful of commercially valued trees, such as teak, sal and deodar. This deprived the tribals and peasants of the forest produce they depended on, and even British experts pleaded that a certain relaxation of the wooden-mindedness of forest departments would greatly help the productivity of Indian agriculture and thereby permit higher levels of taxation (Voelcker 1893; Gadgil and Guha 1992).

The British wanted to retain India as a supplier of cheap raw materials and a market for higher-priced manufactured goods. Industry was therefore discouraged, as was scientific and technical education. A few industrial enclaves did develop regardless, such as the textile mills of Bombay and the Tata iron and steel mill in Jamshedpur. But these enclaves merely remained parasitic and did not serve as foci of modernization. The textile workers lived in miserable tenements in Bombay, leaving their families behind in the countryside, so that no broader, educated working class could emerge. The electric power that fed these mills came from hydroelectric projects built by summarily ejecting peasants, whose lands were submerged without due compensation. Little indigenous technological innovation accompanied this industrial development.

The fluxes of resources between India and Great Britain were highly asymmetrical under this regime. Out of India flowed large quantities of biological produce, rice and cotton, jute and indigo, tea and teak, as well as gold and precious stones. These commodities were produced cheaply; tea plantations in northeastern India, for instance, were set up by taking over tribal lands without any compensation; on them worked labour under conditions approximating slavery. There was little processing of these outgoing biological and mineral produce to add value to them. The return flows were of much smaller material quantities of value-added products of British manufacture.

As rulers, the British assiduously gathered and transmitted back home information on India's landmass, its plant wealth, its people and their customs. The topographic survey of India, the compilation of floras of British India, district gazetteers and ethnographic memoirs consumed a great deal of effort on the part of the colonial rulers (Viswanathan 1984; Desmond 1992). But very little technical information flowed from Britain to India. The Indian elite learnt the English language and read English literature. There was, however, a systematic attempt to keep scientific and technical information from reaching India. When J.N. Tata, one of the pioneers of Indian industry, proposed that an Indian Institute of Science be established to promote indigenous industry, he was consistently discouraged by the British administration. But the institute did come into being in 1909 and in its first decade actively helped establish indigenous technical enterprises. Its council then discouraged interaction between the institute and industry, forcing the faculty to concentrate on basic research. In the event, it accomplished hardly any innovative or original research either (Subbarayappa 1992).

The British omnivores were of course assisted in the task of mobilizing and draining the country's natural resources by representatives of the Indian omnivores of pre-colonial times. Indian society is an agglomeration of tens of thousands of endogamous communities, mostly Hindu caste groups, that observed a hereditary division of labour. In this traditional hierarchy the surplus, primarily of the agricultural produce of the countryside, was appropriated by the three upper strata of Indian society: the *Brahmans* (priests), *Kshatriyas* (warriors) and *Vaisyas* (traders). The two lower strata, *Sudras* (a category which included peasants, herders and fisherfolk, and the more skilled or prestigious service and artisanal groups) and erstwhile untouchables (i.e. the least skilled and prestigious service and artisanal groups), made up the bulk of the population and subsisted as ecosystem people. During colonial rule the upper strata helped serve the British objective of appropriating the surplus of the country's resource production, just as they had earlier collaborated with the many Indian chieftaincies and kingdoms in a similar endeavour, albeit at a less intense rate and on a more limited spatial scale. The task of collecting and transmitting agricultural land tax was assumed principally by the warrior and priestly castes, who became owners of large tracts of land. The trader castes became partners of the British in the task of exporting natural resources and importing and distributing the goods of manufacture. The priestly castes came to man the lower echelons of the bureaucracy that ran the state apparatus.

At this time, the function of the state apparatus was relatively simple. It chiefly consisted in the collection of revenue and the maintenance of law and order. A small number of Indians were selectively educated to assist in these functions, but very few had access to science and technology. The Indian masses, whether peasants, fisherfolk, artisans, nomads or tribals, were left out, with very limited opportunities of education or of jobs in the bureaucracy, in the learned professions or in the slowly developing industrial sector. The tiny organized services and industries sector thus remained the monopoly of upper, literate and trader castes whose traditional divide from the primarily rural lower castes was further reinforced. Meanwhile, the colonial administration remained deliberately alienated from the people at large, with provisions such as the Official Secrets Act shielding it from any public scrutiny. This all-powerful state apparatus bore down heavily on the impoverished peasantry, with the once well-organized village-level systems of management and self-governance largely destroyed. Also gone was the traditional basis of subsistence of a large number of artisanal and service communities. Thus weavers were deliberately forced out of work to ensure a market for British textile mills, while river ferrymen went out of business with the building of bridges and nomadic traders with the expansion of roads and railways. A large populace therefore came to depend on cultivation of land and agricultural labour for subsistence, a populace that began to grow rapidly after the First World War, following the stagnation caused by serious famines and epidemics that characterized the period between 1860 and 1920. All this meant that the traditional network of mutual obligations characterizing Indian rural society was seriously disrupted. Of course, different communities in this network had always formed highly inequitable relationships with each other, but now the fabric itself was greatly strained.

One expects an imperial power to withdraw from a colony when the value of resources usurped is no longer attractive enough to offset the cost of such usurpation. When the British conquered India in the late eighteenth and early nineteenth centuries it had a substantial surplus of agricultural production. This surplus had largely disappeared by 1920. With the spurt in population growth that followed this period, a serious deficit developed. India's forest resources had also been seriously depleted by the Second World War (see Guha 1983), while its mineral resources were not very promising. At the same time the Indian elite had become increasingly conscious of the drain of the country's resources, and come to appreciate the possibilities of diverting these resource fluxes in their own interest. A section of the elite therefore took up the cause of Indian independence, a cause that came to be increasingly supported by the wider population as well (Sarkar 1983). For the British this meant that the costs of usurping the dwindling resource base of their colony were becoming excessively high.

A NATION IN THE MAKING

It was at this low point that the British departed from India in August 1947. State power now passed into the hands of the landowning warrior and priestly castes of the countryside and the priestly and trader castes of the cities. This alliance was committed to the ideal of halting the drain of India's resources abroad. At the same time it was eager to refashion the pattern of resource use to serve its own interests. There were obvious limitations to what could be achieved in the framework of a low-input agrarian economy, one no longer capable of yielding much of a surplus. The solution obviously lay in industrialization; in tapping the energy of coal and petroleum, of hydroelectric power, in producing steel and cement and using the resources so generated to promote manufacture. The way forward also lay in the intensification of agriculture, by irrigating large tracts of land under river valley projects and supplying them with synthetic fertilizer and pesticides. Such a process of development could create a substantial new base of resources whose surplus could support the urban–industrial sector and the rural landowning elite.

This was the model of development India opted for, rejecting the alternative, once offered by Mahatma Gandhi and some of his followers, of crafting an agrarian society of village republics making low levels of demands on the resources of the earth by living close to subsistence (Kumarappa 1938, 1946). The latter path promised no surplus resources that could be channelled to the elite. On the contrary, it called on the central apparatus of the state to surrender its powers in favour of the masses in the countryside. It is possible that the rural masses would have opted in favour of the so-called Gandhian model. Under India's democratic system they had, in theory, an opportunity to do so. But people at the grassroots have so far ended up having very limited influence in India's parliamentary democracy. Under this system politicians are able to go back with impunity on most promises made at the time of election. Once elected they can vigorously pursue their self-interests without hindrance. When political power was quickly monopolized by the rural and urban elite after independence, this elite saw its interests as being far better served by the model of rapid industrial development.

'There is no free lunch in this world' is what the biologist Barry Commoner (1971) once termed the First Law of Ecology. Somebody had to pay for the wholesale intensification of resource use in independent India. In 1947, the state had substantial monetary resources, in part thanks to the contribution made by India in assisting the British in the Second World War. However, the nascent industrial and organized services sector had far fewer resources at its command. Thus the process of intensifying the indigenous use of country's natural resources would have taken off at a rather slow pace if it had had to be paid for only by private enterprise. But there was already a well-established tradition of the British colonial state stepping in to subsidize private enterprise – albeit those units largely under British control. Then the state had readily assigned tribal land to tea plantations and conferred draconian powers over plantation labour to the British estate owners. The state had stepped in to acquire farmland cheaply to set up river valley projects and to quash peasant resistance to such projects. On independence the Indian state elaborated a new model of a mixed economy to take this process of state intervention much further.

This model was inspired in part by Soviet communism. Under Stalin the Soviet state had mopped up the surpluses of the countryside to build heavy industries. This was accompanied by tremendous suffering by the masses of people and degradation of the environment, but at that time the world was little aware of the magnitude of these problems (see Conquest 1968). The Indian political elite under Prime Minister Jawaharlal Nehru held the Soviet Union as the model to be followed, with state-run enterprises taking over the task of producing electricity and steel, fertilizers and radio broadcasts, and running trains and aeroplanes. But unlike in the Soviet Union, private enterprise was allowed to function under state regulation, with freedom to amass private property. Indeed, on the eve of independence leading Indian

Plate 1 Ecological refugees in urban India are engaged in a desperate struggle for a few buckets of water

industrialists not merely supported but actively urged the acceptance of an interventionist role by the state apparatus (Thakurdas *et al.* 1944).

The process of the intensification of resource use in independent India thus became the charge of a bureaucratic apparatus inherited from the British. This was an apparatus fashioned primarily to better organize the drain of resources from the Indian countryside. The British were interested in acquiring these resources as cheaply as possible. They had no interest in the sustainable use of these resources, viewing with little concern destruction of extensive tracts of forests in the vicinity of railway lines or waterlogging of croplands in new irrigation projects (Howard 1940; Whitcombe 1971). This apparatus with its historical baggage was now put to the service of a new set of political masters.

In this framework, the process of development has come to be equated with the channelling of an ever more intense volume of resources, through the intervention of the state apparatus and at the cost of the state exchequer, to subserve the interests of the urban and rural elite. As a result, state subsidies have become a central element of the development process in independent India. These subsidies have served to lower the prices of many goods and services primarily for the more privileged segments of Indian society. They have also absorbed a large fraction of the costs of transporting these goods and services to the centres of concentration of the country's elite.

Plate 2 India's elite now reach out to the remotest parts of the country, as on this island of the Andaman–Nicobar chain

AN ECOSYSTEM PERSPECTIVE ON DEVELOPMENT

The coast

The process of development in free India may be illustrated in terms of fluxes of material, energy and informational resources among the six major ecological regimes and urban centres of the country, as well as the outside world (see Table 1). Until independence the coastal waters of India were fished by artisanal fisherfolk employing rowing boats and sailing boats, with the fish consumed largely along the coastal towns and villages. Since fish was relatively inexpensive, part of the catch was employed as manure, notably in coconut orchards.

The fisheries development effort following independence supported mechanization, with the introduction of trawlers and purse seiners, the development of cold storage and canning facilities and the promotion of export of marine products. Other developments have included the establishment of a series of chemical industries on the coast, for instance a petrochemicals complex at Baroda and a rare-earth plant near Thiruvananthapuram, and the rapid growth of several coastal towns and cities.

All these developments have involved substantial investments of state funds. Thus purchases of trawlers have been heavily subsidized, and Indian Petrochemicals at Baroda is in the public sector. Calcutta has encroached on

Plate 3 Mechanized fisheries and chilling and canning plants today support a major export industry of India

the salt marshes towards the coast with the costs of reclamation almost entirely borne by the government. The mechanization of the fisheries initially enhanced the fish catch by permitting fishing in waters further off the coast, and the development of cold storage and canning has greatly increased the movement of fish and shrimps away from the coast into inland towns as well as for export abroad. Indeed, the export of shrimps has become an important source of foreign exchange for the country. But mechanization, processing and transport also means a greatly increased dependence on petroleum and other external energy and material inputs. Furthermore, the mechanized craft have continued to fish close by the shore, although this zone is supposed to be reserved for countryboats. This has resulted in overfishing and a decline in total fish catch in several states. The extensive coastal pollution has further depressed fish catches, for instance around the chemical industries complex at Baroda. Unable to compete with mechanized craft, fisherfolk have been significantly impoverished. With processing and export the prices of fish have risen sharply so that the poorer segments of the coastal population are now denied access to this important source of protein; nor can fish manure any longer be applied to coconut trees. In summary, the state has invested in the transforming of fisheries so that they bring considerable profit to the traders who have newly entered the industry, through the enhanced supply of fish to urban centres in India and to countries such as Japan and the USA. These effects have been documented by Kurien (1978) and Kurien and Achari (1990).

Inland waters

Fresh waters in India's streams and lakes have long been used for irrigation to enhance agricultural production, and the surplus available to the ruling classes. Evidence of the earliest irrigation works in India dates from the Mauryan Empire two thousand years ago. In south India chieftains promoted the construction of tens of thousands of village tanks in the medieval period (Ludden 1985). The Mughals built canals on the Yamuna. Most of these systems were maintained by local communities, arrangements which continued to some extent through the British period as well. It was under the British, however, that new technologies made possible construction of really large-scale irrigation works. Under the British too, water began to be used for hydroelectric power generation, a major step beyond the traditional water-mills to grind grain that were found in Himalayan villages. However, the colonial rulers had no interest in subsidizing Indian farmers, big or small. While the state stepped in to build dams, it collected irrigation levies that generated adequate levels of returns on the investment. The long-term health of farmlands was, however, ignored, and problems of waterlogging began to plague Punjab and Uttar Pradesh during the period of British rule (Whitcombe 1971; Agnihotri 1993). Power generation was in private hands and (as

Table 1 An ecosystem perspective on development: a summary

				From:			
To:	*Sea*	*Inland waters*	*Forests*	*Grazing lands*	*Farmlands*	*Urban centres*	*Abroad*
Sea	—	—	—	—	—	Sewage, industrial effluent, trawlers	Technology of mechanized fishing, fish processing
Inland waters	—	—	—	—	Pesticides, fertilizers	Sewage, industrial effluent, dams	—
Forests	—	—	—	—	—	Forest-based industry	—
Grazing lands	Chicken feed	—	—	—	—	Dairy industry	—

Farmlands	Fish manure no longer available	Irrigation, water, hydroelectric power	Manure, small timber, fuelwood	Dung as manure	—	Fertilizers, pesticides, high-yielding varieties, green revolution technology	Green revolution technology, diesel
Urban centres	Fish, shrimp	City water supply, hydroelectric power	Fuelwood, timber, paper, polyfibre, tea, spices, milk	Milk, meat	Grain, sugar, cotton, tobacco, bricks	—	Petroleum, electronic goods, technology
Abroad	Shrimp	—	Plywood, polyfibre, textiles, tea, spices	Leather goods	Cotton textiles, crop genetic resources	Trained manpower, software, pharmaceuticals	—

Note: This table depicts the principal material, energy and informational flows within different ecological regions in India and the outside world described in Chapter 3) the British government did help the Parsi industrial house of the Tatas overcome peasant resistance and acquire land cheaply for its hydroelectric projects in the Maharashtra Western Ghats. But the power was sold by the Tatas to Bombay consumers at rates high enough to bring in a tidy profit. However, freshwater fishing was left largely alone under the British regime and problems of water pollution were modest given the comparatively low levels of urbanization and industrial development.

Following independence, the mobilization of water resources was seen as the key to stepping up agricultural productivity as well as the supply of electrical power. Water was also needed to service the rapidly growing urban centres. As urban centres grew and industries developed, water also had to serve to disperse the waste products. Even in the worst years of drought the Indian landmass receives on average 6,000 mm of rain; in years of good rain this climbs to 10,200 mm. Including the water received from the Himalayan watershed this comes to an average of 4,200 billion m^3 for the country as a whole. Given a population of 840 million people, this translates to nearly 5,000 m^3 per head. These are reasonably adequate levels; but in India this precipitation is largely confined to three to five months of the monsoon season. As a result serious water scarcities can and indeed do develop in the dry season even in a town like Cherapunjee in Meghalaya, which in an average year receives as much as 10,869 mm of rain (Centre for Science and Environment 1987).

Across India water must be impounded, preferably at a height, and transported as and when required to the point of use, whether for irrigation, industrial or domestic use, or power generation. In this old and densely settled country all such activity is bound to deprive some people of access to water or land, while improving resource access for others. For example, the westward diversion of the waters of the River Koyna from near its origin in the Western Ghats to a power station on the west coast has meant lower water supplies for villages lining the bank of the river on its eastward course towards the Deccan plateau. Other villages, mostly populated by small peasants and buffalo herders, were submerged under the dam, the inhabitants thus deprived of the only basis of subsistence they knew of. Subsequently, local people have also suffered from earthquakes triggered off by the impounding of this large body of water. Meanwhile, the power generated from the Koyna hydroelectric project has primarily fed industrial development in distant Bombay (Paranjpye 1981).

Other dams in the Maharashtra Western Ghats have taken water eastwards to irrigate lands (mostly under sugarcane), or to industry and cities. They have nurtured pockets of prosperity while at the same time depriving others of their livelihood. Such projects have also created access to hilly forested areas till then left untouched, thereby triggering deforestation. All these costs have been borne by small peasants, tribals, herders and rural artisans, who have been most inadequately compensated for the losses they have suffered

Plate 4 River valley projects opening up access to remote areas have triggered deforestation in many parts of India

(Gadgil 1979; Sharma and Sharma 1981). Also ignored are the longer-term costs: for instance, the possibility of earthquakes triggered some decades later by the building up of strains in the geological substrate (Valdiya 1993). The benefits of large dams have gone primarily to the industrialists of Bombay, the sugar barons of western Maharashtra, and urban dwellers in general, who have received power and water at costs far below those incurred by the state exchequer (Singh 1994).

While promoting the construction of large dams, the state apparatus has overseen the collapse of traditional systems of smaller village tanks and distributaries. During the British regime these community-based management systems were sometimes permitted to continue, for they were efficient and made possible collection of land revenue at higher levels. In independent India, the state has no longer considered tax on agricultural land to be an important source of revenue. Instead, the state apparatus has wanted to enhance its control over the resource base. So all over India village tanks have been taken over by the Minor Irrigation Department, leading to a decay of local management systems and the rapid siltation of the tanks themselves (Somashekara Reddy 1988; Shankari 1991; Bandyopadhyay 1987).

The easiest solution to pollution is its dilution with the help of water and air, public goods in theory available to all. This is the solution adopted everywhere in India as growing industrial activity, exploding urban concentrations and the growing use of agrochemicals in intensive agriculture all pump in more and more of polluting substances into India's environment (Alvares 1992). By and large little investment, private or public, has gone into treating industrial effluent or city sewage, and none at all into checking the discharge of agrochemicals. But this has not been without its attendant costs. These include the deprivation of drinking water to people and cattle dependent on natural watercourses, the loss of sources of irrigation water, a decline in fisheries, and, in some cases, even the poisoning of the underground aquifer. These costs have been widely dispersed, mostly borne by the ecosystem people, who must look after their own water needs. The benefits have accrued to those in the organized services–industries sector, the city dwellers, the better-off cultivators using fertilizers and pesticides. Here the pattern is very similar to that of the processes of exploitation of the resources of the sea. The state steps in to enhance resource availability to omnivores. If in the process the resource base of ecosystem people is attenuated or destroyed through pollution, the state barely intervenes. It certainly does not compel omnivores to bear these costs.

Plate 5 Fly ash blankets the land in the vicinity of coal-fired power plants in many parts of the country

Forests

The exploitation of India's forests provides a striking example of how state policies have favoured omnivores at the cost of ecosystem people, while at the same time promoting the exhaustive use of a renewable resource. The colonial state had already laid the foundation for this, by taking over land as reserved forests dedicated to the supply of cheap timber to build teak ships, to lay down railway lines, to put up British cantonments and provide for the two world wars. Apart from mills to saw timber for urban housing and furniture, little forest-based industry developed during British rule. These were, of course, only one side of the demands on forest produce, for the masses of India's ecosystem people needed large quantities of fuelwood, small timber and thatch. They built huts out of bamboo slats and stored grain in bamboo baskets. Their agricultural, fishing and hunting implements were all fabricated from plant material. Their livestock grazed extensively in the forest and in some parts tree fodder lopped by hand was important for their maintenance.

But the Indian Forest Department, founded during British rule, and in effect India's single largest landlord, viewed all these needs of the ecosystem people as a burden, as 'biotic' not 'anthropic' pressure, as if the people behind these demands were less than human. (Indeed, forest working plans classified 'man' as one of the 'enemies' of the forest.) Reluctantly some lands were set aside, as revenue 'wastelands', from which ecosystem people were expected to meet their substantial and vital biomass needs. However, ecosystem people had no longer any rights over these lands, only 'privileges', so that all these areas became no man's lands, overused without restraint by all and sundry (Gadgil and Guha 1992).

This framework has changed but little since independence, and such changes as have occurred have come about only in the past few years (see Chapter 7). At the same time the demands for forest resources have risen tremendously as the domestic consumption of paper, polyfibre yarn, plywood, etc. has soared, along with the continuing rural demand for fuelwood, fodder and small timber. Forest resources have been made available for industrial use at throwaway prices: bamboo at Rs 1.50 per tonne, when the prevailing prices were Rs 3,000 per tonne in 1960; at Rs 600 per tonne after much pressure, when the prevailing prices are Rs 12,000 per tonne. Indeed, it was the guaranteed supply of pine resin to a factory in Bareily at subsidized rates, while the locally set up cottage industry received erratic supplies at higher rates, that prompted the Sarvodaya workers of the Alakananda valley to question the working of the Forest Department in the late 1960s. Soon after this came the award of rights to fell trees, valued locally as a source of agricultural implements, to a sports goods factory in distant Allahabad. This precipitated the famous Chipko *andolan* in March 1973, with unlettered peasants threatening to hug trees to prevent them from being cut (Gadgil and Guha 1992; Gadgil 1989; Guha 1989a).

With the debate generated in the aftermath of Chipko, there has come about a shift in the way India's forest resources are being managed. This has included an official acknowledgement, in the National Forest Policy of 1988, that the biomass needs of ecosystem people must have primacy over the commercial demands of omnivores. Some attempts have also been initiated to set up management systems involving local communities so that no man's lands could be brought under a more regulated regime of harvests (Agarwal and Narain 1990; Malhotra and Poffenberger 1989; and for a detailed assessment, see Chapter 7 of this book). But for the most part these policy initiatives have not translated into practice, with omnivores stonewalling all attempts at sharing power with ecosystem people. In the meantime the subsidized supply of forest raw material to industries goes on, while the large masses of ecosystem people continue to meet their biomass needs in a completely unregulated fashion from open-access lands. Thus the growing numbers of ecosystem people and the growing appetite of omnivores impose larger and larger demands on a dwindling resource base, contributing to widespread deforestation. This also means a serious impairment of the ecosystem services of the forests; in soil and water conservation, as a repository of biodiversity and as a means of sequestering carbon (Gadgil 1989).

Grazing lands

Cattle and sheep, and to a lesser extent buffaloes, goats and pigs, have been major elements of India's rural economy. Grazing on the stubble and straw of crops and on the scrub and forests surrounding villages, they have been a very significant source of nutrients to replenish the poor soils that support much of Indian agriculture. Milk and milk products have been important components of Indian nutrition, and bullock power is critical to farm operations and transport. Traditionally, livestock was maintained by a majority of farmers as part of a mixed agriculture–animal husbandry system, as well as by specialist herders who moved with their cattle or sheep over large distances during the dry season. The farmers grazed their animals on village common lands subject to community control; the nomadic herders too had their own understanding as to the areas to be grazed by different groups and the relationships to be maintained with settled cultivators (Whyte 1968).

A prime focus of the British was on enhancing the revenue from cultivated lands; grazing lands were to them wastelands preferably to be brought under cultivation or converted into reserved forests. Irrigation was used as an important device to achieve the former objective – thus large tracts of grazing lands in Punjab were brought under the plough (Agnihotri 1993). Traditions of grazing cattle in the forests, as by nomadic Gujjars in the western Himalaya, were severely restricted by the constitution of reserved forests. Village grazing lands were treated as revenue wastelands converted into no man's lands subject to overuse.

This maltreatment of the grazing requirements of India's large livestock population has continued after independence. In particular, the roles of livestock in maintaining the fertility of farmlands and as a source of power for tilling and rural transport have been largely neglected. The emphasis has instead been on the use of synthetic fertilizers and the mechanization of agricultural operations and rural transport. The omnivores have, however, been interested in dairy products, in meat – especially of goat and sheep – and in leather goods. State interventions have therefore focused on the more efficient channelling of dairy products, meat and leather to the cities; and leather and leather products abroad. Substantial investments have gone into transporting, storing and processing milk, so that today the city of Bombay can draw its milk not just from much of its own state of Maharashtra but also from Karnataka, Gujarat and Madhya Pradesh as well. This of course implies much larger demands on petroleum and other energy sources. The goat population has increased at rates far exceeding those of other livestock as consumption of goat meat by the growing numbers of omnivores has been on the rise. But this has not been accompanied by attempts at enhancing the availability of fodder. Indeed, the efforts at enhancing agricultural production have invariably led to the introduction of high-yielding varieties with shorter stature and hence to a lower production of straw relative to grain. Nevertheless, the overall increases in productivity in tracts of high-input agriculture have been sufficient to offset this and to increase the total production of straw in these areas. The net result has been a greater dependence on crop residues for fodder, with the establishment of dairy industries in tracts of irrigated agriculture, while in other parts forests, scrublands and grazing lands are being increasingly overgrazed and degraded. As this process of degradation proceeds, the landless and poorer peasants shift to maintaining goats in preference to buffaloes or cattle, thus intensifying grazing pressure on the degrading vegetation (Gadgil, Pillai and Sinha 1989; Gadgil and Malhotra 1982).

With all this, the total livestock population of the country continues to go up, yielding increasing quantities of leather. These leather goods have become a very significant component of the country's export trade. Leather production is accompanied, however, by a serious problem of pollution from tanneries, for example in the cities of Madras and Kanpur (Alvares 1992). Little investment has gone into correcting this problem; and again the cost is borne by poorer people dependent on water and fish from streams and rivers. To sum up, then, state interventions to organize dairies, to provide cattle and goat purchase loans, or to promote leather exports have primarily helped to enhance the availability of dairy products and meat for city dwellers, to generate income for farmers with irrigated lands and to bring in foreign exchange, all at the cost of the degradation of India's forests, grasslands and water bodies.

Farmlands

A simple approach characterized the British management of Indian agriculture. The British encouraged the bringing of increasing amounts of land under cultivation and under irrigation so as to extract as much revenue as possible from land cess. They also encouraged the production of crops like cotton and indigo to provide cheap raw material for British manufacture. In the process, the Indian peasant lost control over land to large, often absentee, landlords, and was perpetually in debt. Agricultural productivity remained largely stagnant and there were several disastrous famines. Small irrigation systems remained functional under community management and were supplemented by a few larger systems that enhanced agricultural productivity, for example in the dry tracts of Punjab. These large-scale dams were accompanied by problems of displacement of graziers, of waterlogging and salination of cultivated lands, and the outbreak of malaria (Whitcombe 1971, 1982, 1993).

State objectives in the farm sector shifted substantially after independence. A high priority was to enhance food production, along with the production of raw material for agro-based industries such as textiles, sugar and cigarettes. There was also a clamour from the peasantry for relief from debt and for land for the tiller. There were two sets of options open to the Indian state to meet these objectives. First, it could either spread inputs, enhancing agricultural productivity over wide tracts, or furnish concentrated inputs to restricted tracts. Second, it could either push through land reforms, or permit the continuation of large land holdings and share cropping.

In both cases the first of the two options would have favoured the large masses of ecosystem people, the second option the omnivores. But as it happens, by and large options favourable to omnivores have been followed over much of the country. Thus the strategy of the green revolution has been to pump in water, fertilizer, pesticides and high-yielding varieties to selected areas. It is this strategy that has successfully raised the productivity of Indian agriculture over the last quarter of a century. This success has taken the pressure off the need to enhance productivity on a broader base, a path that would have called for land reform. Since independence, the pressure for land reform has come both from Gandhians and from the political left of various hues ranging from Maoists reposing their faith in armed struggle, to the two mainline communist parties, the Communist Party of India (CPI) and Communist Party of India (Marxist) (CPM), which work within the framework of parliamentary democracy. Gandhian efforts have had little long-term impact, and land reform has gone furthest in West Bengal and Kerala, states with long years of communist rule. It has made least progress in the so-called BIMARU states: Bihar, Madhya Pradesh, Rajasthan and Uttar Pradesh (see Herring 1983; Joshi 1975; Jeffrey 1992).

The strategy of pumping in concentrated inputs in limited areas could never have been implemented if the farmers who benefited had to pay fully for inputs. Instead, the state has stepped in to massively subsidize water, power, fertilizer and pesticides. Ecologically speaking there are serious dangers in such an approach: dangers of waterlogging, of salination, of overuse and nitrate pollution of ground water, of the concentration of pesticides in the environment and the decimation of many non-target organisms, of a build-up of resistance in pest populations and thus of large-scale pest outbreaks in monoculture crops (Shiva *et al.* 1991a). Indeed, all these problems have materialized, along with problems of highly inefficient use of the subsidized resources. But in the process not only the chosen farmers, but those generating and directing the subsidized outputs – industry, bureaucrats and politicians – have done well. Moreover, concentrated inputs in restricted areas more easily generate surpluses that can be skimmed off to urban areas. All these subsidies permit the prices of agricultural produce to be kept at low levels. Further subsidies in food supply, primarily available in urban areas, have helped cope with the rapid growth in city populations.

The enhancement in the productivity of Indian agriculture is now shown to have been accompanied by a very low efficiency of resource use. As a result, while productivity per hectare has gone up substantially, productivity per unit of external energy input (for instance, in bringing water to the field, in manufacturing fertilizers, pesticides and so on) has sharply declined (Department of Science and Technology 1990). Especially since India has only a limited availability of petroleum, this has meant a greater dependence on imports. Foreign exchange must be earned to pay for such imports. Here textiles, along with tea, fish and leather, are critical to India's export earnings. The yarn for these textiles comes partly from polyfibre out of eucalyptus plantations that have replaced large tracts of India's natural forests, and partly from the cultivation of cotton. Indeed, there is an aggressive state-sponsored drive to enhance cotton production. But cotton, of all crops, demands very heavy doses of pesticides with all the attendant risks of environmental damage.

Farming in India has traditionally been a system of a rich mix of varieties of cereals and legumes. While wheat and millets and possibly rice were introduced to India from outside, a number of legumes, for example mung and pigeon pea, were domesticated there. Intercropping and rotation of cereals, nitrogen-fixing legumes and the extensive use of cattle dung as a manure were the key to the sustainability of traditional Indian agriculture. The cropping pattern also involved the use of thousands of locally adapted varieties. Modern intensification has destroyed this diversity – its emphasis is the creation, instead, of homogeneous stands of crop varieties that can perform well only if supplied with large amounts of water and heavy doses of nitrogen, potassium and phosphorus and protected by intensive application of weedkillers, insecticides and fungicides. Such systems have been perfected only in the case of a few cereal crops such as wheat, rice and sorghum. In particular, legumes, a most characteristic feature of Indian agriculture, are not a part of these systems of production characterized by high inputs and low diversity. Production of legumes in India has therefore remained more or less stagnant, even as cereal production has been growing along with human population (Bajaj 1982). The result is a great scarcity of pulses, a sharp rise in their prices and protein deprivation for the poor. A rise in the price of fish as this is diverted to the tables of omnivores in India and abroad of course further aggravates this problem.

While the rich diversity of India's crop varieties has been vanishing from the fields of the farmers, it has at least in part been preserved in agricultural research stations in India and abroad, and in seed banks owned by commercial companies. Much of this diversity is now under the control of foreign agencies that gained access to it entirely free of charge under a regime proclaiming it to be the common heritage of mankind. Indeed, India's crop genetic diversity has already contributed greatly to world-wide cereal production. Some years ago, for instance, an insect pest called the brown leafhopper inflicted billions of dollars' worth of damage on the rice crop of southeast Asia. A solution was finally found in the form of a gene conferring resistance to this pest in a rice variety from Pattambi in Kerala, located in the collection of the International Rice Research Institute in the Philippines. This Pattambi gene was obviously worth a great deal but there was no question of any payment under the common-heritage regime. This regime is now being scrapped under pressure from multinational seed companies based in the West, which see tremendous commercial potential in the exploitation of genetic diversity, especially with the advent of modern biotechnology. India has thus exported entirely free of charge a tremendous resource base built up through the efforts of millions of peasant households over thousands of years. It will now have to pay heavily for the new varieties incorporating these genes that the multinational seed companies are likely to produce in coming decades (Shiva *et al.* 1991a).

The developments in India's farm sector may then be summarized as a selective enhancement of agricultural productivity, which has augmented the agricultural surplus at the disposal of the omnivores while increasing the need for external inputs, especially those based on petroleum. These developments have left the bulk of India's peasantry, the dryland farmers and other weaker sections no better off than before as the prices of pulses, their principal source of protein, have skyrocketed. At the same time, the pumping in of large quantities of water and agrochemicals in selected areas has resulted in manifold environmental problems (Alvares 1992).

Urban industrial enclaves

People obtain a subsistence in very many different ways. As hunter-gatherers, fisherfolk or miners they gather natural resources, partly for self-consumption, partly for exchange. As cultivators or herders they husband crops or livestock. As artisans or employees of an industrial corporation they process a variety of resources to add value to them. As shopkeepers or employees of a trading concern they arrange to distribute resources. As barbers, teachers, priests, bureaucrats, policemen, chieftains and politicians they provide a range of services. The variety of ways in which people thus subsist has multiplied and grown more complicated with technological advances. At the same time the terms of exchange have come more and more to favour activities adding value to resources in the organized industries and services sector in relation to those which simply gather or husband resources.

Britain gained control over India at a stage when the British were forging ahead in the development of technologies that could add value to resources. Their interest, therefore, lay in establishing a relationship with the colony where they could supply value-added commodities in exchange for biological and mineral resources gathered or husbanded within India. This in part required the active suppression of competition, for example of Indian shipbuilders or weavers, who had the competence to produce superior value-added products. In part, such a relationship was secured by ensuring that the new techniques percolated to India very selectively. Telegraph and railway lines were of course laid down in India to promote resource flows, but railway engines and telegraph machines continued to be manufactured abroad (Sangwan 1991). The introduction of these modern communication and transport devices meant that at a stroke a number of activities pursued by a substantial population of pre-colonial India became obsolete. These included the occupations of river ferrymen, itinerant traders and nomadic entertainers. Also suppressed in large measure were activities like shifting cultivation (Gadgil and Guha 1992). During British colonial rule therefore there was a shrinkage in the range of occupations that Indians could pursue, leading to a mounting pressure on cultivated lands. Since the productivity of agriculture grew only at a snail's pace, while population growth gathered pace after the First World War, there were large masses of underemployed people on the eve of independence.

It was in this setting that India began its own drive to industrialize, i.e. to add value to resources on its own, instead of merely exporting raw materials. In this setting labour was abundant, human-made capital, i.e. artefacts employed to add value to resources, scarce. With a large population and a base of natural resources that had been considerably depleted, the country was not especially endowed by nature either. It was clearly by no means easy for industrialization to take off. For it to gather pace industries needed to make profit, which could come from three routes: (a) by charging high prices, which would be possible if competition was suppressed and a sellers' market created; (b) by getting cheap access to power, water, land and raw materials, which would only be possible if the state stepped in to subsidize the cost of these resources which were by no means abundant, and indeed under considerable demand by a large population; (c) by taking advantage of cheap labour, since huge numbers were unemployed.

The soft options were evidently to get the state to create a sheltered market and permit the charging of higher prices than could stand international competition, as well as the passing on of a substantial fraction of the cost of the resources used to the state exchequer. The industrial sector vigorously canvassed for and successfully persuaded the state apparatus to implement such policies. Once such support was assured, it was easiest to go in for capital- rather than labour-intensive technologies, and to concentrate industries in places where the flow of subsidized resources had been well organized. So the possible option of labour-intensive industrial activities dispersed through the countryside, an option that would have been more favourable to the ecosystem people, was never pursued seriously. In deference to the heritage of Mahatma Gandhi, some support did go to village handicrafts, but that too was only in the form of government subsidies, and has slowly eroded over the years (Jain 1983).

The pattern of industrialization in India has therefore been one of a concentration of capital-intensive activities in relatively few areas. These centres, cities like Bombay, Baroda, Madras, Calcutta and Coimbatore, have in consequence been the loci of enormous government investments for

Plate 6 Railway line overhanging the landslide it has caused. The extensive network of railways and roads in the country has often been a source of environmental damage

providing water, power, transport and communication facilities as well as subsidized food supplies. Indeed, in most cases the larger the city, the larger is the quantum of state subsidy flowing in (Maitra 1992). Naturally enough, the country's omnivores are concentrated in such centres. Even the rich landowners whose primary base is the countryside often have another base in the pampered urban areas.

In this scenario, it pays industry to manipulate and bribe politicians and bureaucrats, rather than to worry about technological innovation, efficient resource use or pollution control. So the soft options have been to import technologies and not worry at all about how efficient or environment-friendly these technologies are or were. The inevitable result is that (a) Indian industry is highly inefficient; (b) industry has no commitment to sustainable resource use or the maintenance of a clean environment; (c) employment in industry has grown very tardily. These are the hallmarks of India's high cost – low quality economy.

Plate 7 Untreated sewage from the city of Bangalore poses a serious health hazard for villagers living downstream

This complex of processes favouring a narrow elite of omnivores at the cost of masses of ecosystem people has created, too, large numbers of ecological refugees in the hinterlands. These have flocked to the cities, to which the state has channelled resources, from all over the country. Of course, ecological refugees cannot become full partners in sharing this largesse. As Eduardo Galeano (1989) has written of their counterparts in Latin America, the ecological refugees in Indian cities 'sell newspapers they cannot read, sew clothes they cannot wear, polish cars they will never own and construct buildings where they will never live'. But they may still be better off in city slums than in declining villages, with marginal access to water and to opportunities of employment, at least in the unorganized sector as domestic servants, hawkers and recyclers of garbage. Or, as in the outskirts of Calcutta, they may learn to cultivate vegetables very efficiently using solid organic wastes and city sewage. In consequence city populations have grown rapidly, not really through growth of economically productive activities, but as parasites subsidized at the cost of the hinterlands.

Plate 8 Shantytowns of ecological refugees are an inescapable feature of urban India

This, then, is the current social and environmental scenario in India. But such a parasitic development has its natural limits. Over the years the state apparatus has grown to levels at which it no longer has the capacity to subsidize the omnivores outside its fold; most of its resources are now being swallowed by the salaries and perks of government employees and their political bosses. But with no resources left over to patronize other omnivores, or to pacify ecosystem people and ecological refugees, the state apparatus is in trouble. The only recourse left to it is to beg for resources from abroad. But the global omnivores – represented by the World Bank, the International Monetary Fund and transnational corporations – would not make such resources available for nothing. Their interest is to ensure that Indian markets are opened up to their products. Under pressure the Indian state has little option but to concede these demands. Thus the last few years have exposed India's high-cost, low-quality economy to traumatic competition from abroad. As a result of international competition Indians are being alerted for the first time to the massively inefficient process of resource use that has characterized the country's development process. At the same time the protests of ecosystem people and ecological refugees, who have suffered most over the years, are becoming increasingly evident. To the liberal economist's clamour for *efficiency* these protests have counterposed the equally compelling slogans of *ecology* and *equity*.

It is, then, a propitious occasion for us to examine where Indian society and environment is, how it got there, and where it may proceed. It is our purpose to explore these processes further in the succeeding chapters of this book.

2

PASSING ON THE COSTS

ISLANDS OF PROSPERITY, OCEANS OF POVERTY

In nature, material and energy tend to flow down the gradient. When an iron rod is heated at one end and cooled at the other, heat tends to diffuse from the hotter end. When some coloured water is poured into a bucket full of plain water, the colour gradually spreads throughout the water body. These processes tend to reduce differences, to favour homogeneity. The processes of material and energy fluxes outlined in Chapter 1 do just the opposite. Once a pocket of irrigated cultivation is created, into it flow subsidized fertilizers, pesticides and electric power. Once a pocket of industrialization takes root, it likewise attracts subsidized water, power, raw materials and communication facilities. Significantly, these fluxes tend to be at the cost of the hinterlands. Thus people in the catchment of a dam are legally prevented from using the reservoir water for irrigating their own lands. Refugees from the Koyna dam, which supplies power to the cities of western Maharashtra, cannot themselves afford even kerosene to light their huts at night.

Forty years of planned development has created an India in which islands of prosperity peep out of a sea of poverty. The omnivores inhabiting these islands are securely on firm ground. The bulk of India's ecosystem people are submerged in the sea of poverty. The ecological refugees are hangers on at the edges of the islands of prosperity, somewhat like mud-skipper fishes hopping around on the muddy beaches fringing mangrove islands. From time to time the tide swallows them; they manage to clamber back on to the mud, but can never make it to dry land.

In a democratic society a small minority ought not in theory be able to thrive at the cost of a much larger majority. It does so in India, as in many other countries of the world, by perverting the spirit and subverting the workings of democracy, by managing to concentrate a great deal of power in its own hands. To accomplish this the omnivores have built up an alliance akin to an iron triangle – an alliance of those favoured by the state (industry, rich farmers and city dwellers); those who decide on the size and scale of these favours (politicians); and those who implement their delivery (bureaucrats and technocrats). To illustrate the working of this alliance, take the creation of Salt Lake City, the fashionable new extension of Calcutta. This was once a salt marsh, the northward stretch of the mangrove swamps of the Sundarbans. Enormous investments were needed to drain it, to create dry habitable land and to bring in sweet water. The state made all these investments and created prime real estate abutting the metropolis of Calcutta. The land was then sold to the politically influential and to well-paid bureaucrats at less than 10 per cent of the market price – never mind that the market price itself did not reflect the prior state investments in its reclamation and development. Part of what was so acquired was used by the beneficiaries personally; the rest was sold at huge profits. The land awardees undoubtedly kicked back part of the profits to the politicians and bureaucrats making the awards.

Analogous processes operate all the time: when alignments of new roads and irrigation canals are decided on; when forest raw material is assigned to the paper or plywood industry; when licences to import television sets or to manufacture cars are awarded.

Figure 1 illustrates the operation of this alliance of omnivores, the iron triangle. The prime beneficiaries of this system of state-sponsored resource capture are those in the organized industry and services sectors and the larger landowners in areas of intensive agriculture and horticulture. The state absorbs a large fraction of the costs of water, power, raw material, fertilizers, etc. provided to industry, services and intensive agriculture; it also organizes provision of these resources to the homes of omnivores, again with large subsidies from the state exchequer. This resource capture by omnivores is at the cost of the other five-sixths of the population: the landless labourers, small peasants, rural artisans, herders, countryboat fisherfolk, nomads and tribals; and such of these as have ended up in urban shantytowns.

The masses are persuaded to accept this system through three kinds of devices: by permitting a trickle of handouts to reach them; by ensuring that they remain assetless and uneducated; and, finally, by more active coercion. There is a whole range of rural development programmes that aim to take state-mediated benefits to the ecosystem people; so many indeed that they come to be known only by their acronyms or abbreviations – RLEGP, TRYSEM, IRDP, DPAP and what have you. Some offer straightforward patronage such as large-scale distribution of *dhotis* and saris to the poor. Others provide subsidies for digging a well or purchasing a cow or a goat. Still others generate employment in constructing roads or digging trenches around forest plantations. A portion of the handout does reach the intended beneficiaries, but a substantial fraction is misappropriated by the political-bureaucratic machinery. As a result the programmes often end up unproductively; the farmer may pocket merely half the subsidy due for purchasing a milch cow, the veterinarian the other half for certifying the existence of the mythical cow. There may be false muster rolls of payment for carrying out soil conservation

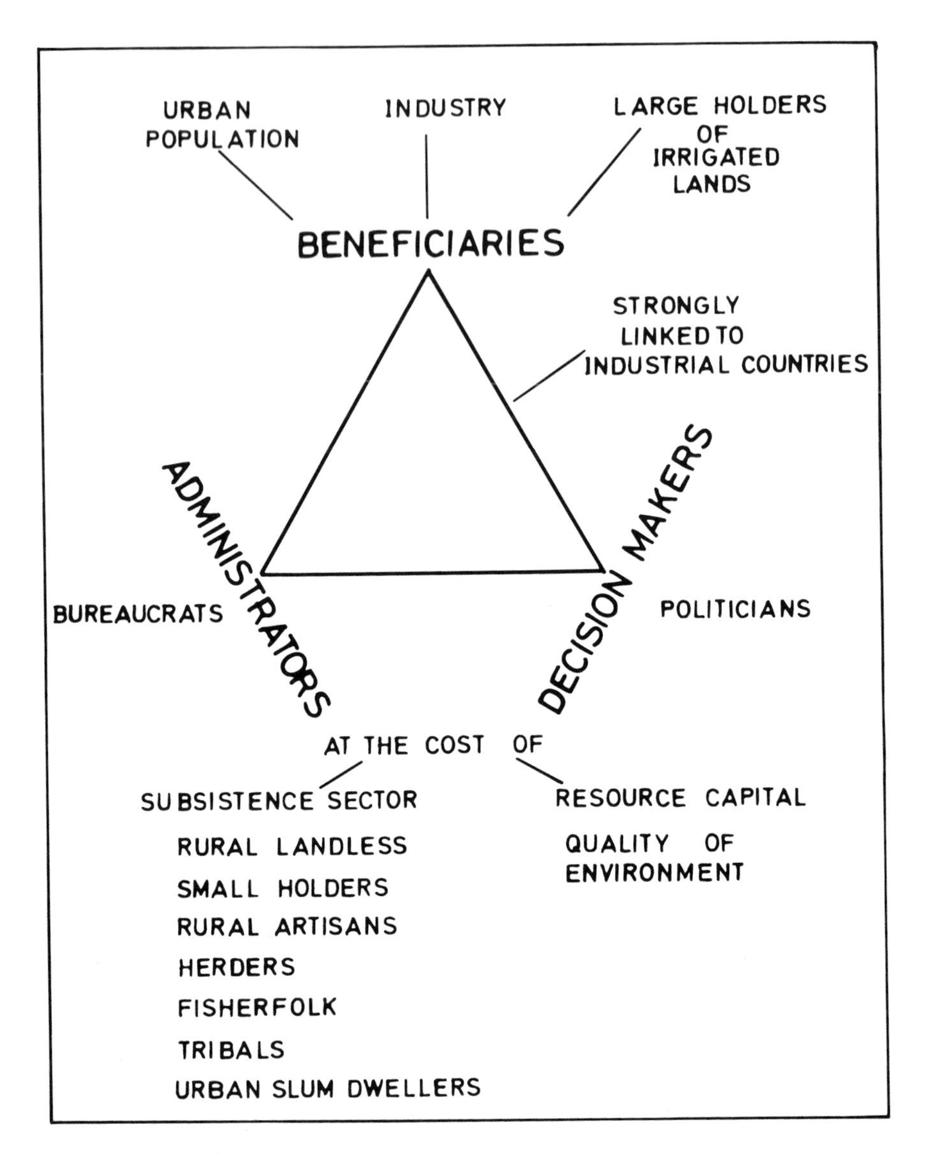

Figure 1 The iron triangle governing resource use patterns in India. Large state-sponsored subsidies have created an iron triangle of components of Indian society benefiting from, administering and deciding upon state patronage. Constituents of this iron triangle are forcing the country into a pattern of exhaustive resource use at the expense of the environment and a majority of the people

work, itself never adequately implemented in the field. Nevertheless, all citizens of India, poor and rich, have come to look upon the state as a source of free handouts. All are involved in a scramble for a share – minuscule, moderate or large – of these handouts, with few questioning the pernicious system itself.

Through forty-five years of independence the masses of people have largely continued to remain assetless. The most significant asset in a predominantly agricultural country is land. Land reform, assigning ownership of the land to the tiller, has made hardly any progress over large parts of India. This has ensured that the poor cannot effectively make themselves felt in the political process over much of the country. It has also meant that efforts at enhancement of agricultural production have concentrated on very limited tracts of land, with a wide range of undesirable environmental and social consequences.

After forty-five years of independence too a majority of Indian citizens remain illiterate. Little money proportionately has been made available for primary education, although quite substantial amounts are invested on training engineers and chemists who permanently settle abroad. Again, the money actually made available is ill spent, since very many teachers do not attend to teaching in more remote villages, but simply draw their salaries. The villagers have no control over teachers, who report only to some education officer at the district headquarters.

However, there is every reason to believe that a very substantial proportion of India's rural people are motivated to absorb education if an opportunity presents itself. Such an opportunity arose in the form of the Literacy Mission, a programme to achieve minimal levels of literacy for all Indians that was initiated by the Government of India in 1988. For the first time this mission seriously attempted to involve the voluntary sector in a major national effort. It was rewarded by an excellent response, especially in a number of south Indian districts. The voluntary sector became organized and began to play an active role in bringing in more accountability to the development effort. Likewise, in the state of Karnataka, teacher attendance in village schools dramatically improved when local-level political institutions, known as *mandal panchayats* and *zilla parishats*, were given a measure of real autonomy between 1986 and 1990. However, these developments are evidently not to the liking of omnivores. Thus the *panchayats* and *parishats* were abolished when the Congress Party returned to power in Karnataka in 1989, even as the involvement of the voluntary sector in the National Literacy Mission has been watered down.

While the rural masses have been cheated of access to education on a broad scale, there have been attempts to buy their acquiescence by reserving a fraction of seats in higher education to the scheduled castes, scheduled tribes and some of the other backward castes. In the absence of serious land reform the assetless have no chance of acquiring good-sized parcels of irrigated land, but through such reservation they do have a slender chance of acquiring an education and a job in the organized industry–services sector. Higher education is the one avenue available for outsiders to join the ranks of omnivores. There has therefore been considerable political pressure to extend reservations for backward communities not only in institutions of higher education, but in the form of job quotas in the state (or omnivore) sector as well. Such reservation opportunities mean that a small number of ecosystem people and ecological refugees can make the transition into omnivory, but without in any way affecting the overall distribution of power and resources between the three classes.

DEMOCRACY AND POWER

Indians can justifiably be proud of being citizens of the world's largest democracy. Except for a brief interlude between 1975 and 1977, there have been reasonably free and fair elections to the state legislatures and national parliament over a forty-five-year period. But this does not mean that the interests of the masses have really prevailed. Since these interests are far more congruent with those of a healthy environment, their subversion by omnivores has often resulted in environmental degradation.

Omnivores have subverted democratic processes on a variety of scales and in a variety of ways. The British initiated and the Indian state continues the assumption of draconian power in the acquisition of land for ostensibly public purposes. These powers, embodied in the Land Acquisition Act of 1894, have been misused to drive peasants off their lands without adequate notice and compensation in order to facilitate a variety of development projects. As early as 1921 the protest against the Mulshi hydroelectric project in western India was precipitated when a British engineer organized the digging of trenches in private fields even before any notice of intention to acquire land had been issued (see Chapter 3 for details). To this day farmers whose lands are being submerged by dams on the River Narmada are protesting that they have not been served due notice, nor have provisions been made for proper rehabilitation following their eviction.

Indian society has had rich traditions of decentralized governance, not least in the sphere of natural resource management. Down the centuries, while rulers came and went, taxing and plundering, India's villages managed their own lands, waters and forests. Recent historical research suggests that pre-colonial community-based management systems had elements of equitable sharing and democratic decision making, and on the whole functioned quite effectively (Guha 1989a; Sengupta 1985). The British dismantled the community management of farmland (where it existed) to facilitate taxation, and took over large areas of community forests as state property. However, they largely left intact the community management of small-scale irrigation works, which facilitated the collection of land tax at higher rates. The Gandhian ideal of reviving India as a land of village republics called for the continuance, and indeed revival, of such community-based management practices (see Kumarappa 1938). Instead, after independence the heavy hand of bureaucracy came down on all such remaining systems, proceeding to dismantle them.

The peninsular Indian Archaean crystalline shield areas of Andhra Pradesh, Karnataka and Tamil Nadu have an intricate network of tens of thousands of tanks constructed over hundreds of years. Over the centuries village communities have maintained these systems through regular desilting and the repair of canals through community labour, the observance of a variety of regulations relating to the maintenance of the catchment areas and the sharing of water which might include assignment of plots for cultivation depending on the availability of water. Use of water from such tanks was regulated by a village functionary authorized and paid for by the community. The systems continued through the period of British rule, although village communities themselves did drastically change through concentration of land in the hands of a small number of people.

On independence it would have been in the democratic spirit to have continued these decentralized management systems, perhaps correcting for excessive domination by large landholders. Instead, the machinery of the Minor Irrigation Department took over the management of virtually all these tanks. This is consistent with the overall drive of the bureaucracy in independent India to take on more and more responsibilities and to enlarge bureaucrats' power and job opportunities, regardless of whether the assigned responsibility can indeed be discharged effectively by them. The experience of the past four decades has been that tens of thousands of tanks so taken over by the state apparatus have almost all fallen into disrepair. The village communities have stopped contributing voluntary labour, while the state is quite incapable of paying for and efficiently executing the functions earlier performed voluntarily. Further, the subsidized supply of synthetic fertilizers, irrigation pumpsets and electrical power has reduced the interest of better-off agriculturists in the silt and water from the tank. So the irrigation authorities are now proposing simply to write off this huge and ancient network of tanks, while urging that the state instead go on constructing newer, bigger irrigation projects (Shankari 1991; Sengupta 1985; Singh 1994; Von Oppen and Subba Rao 1980).

Not all community-based tank management has yet been abandoned and in a number of villages these systems continue, albeit without the co-operation of the official machinery. A proportion of these community-based tanks have also fallen into disrepair, but a significant number of them continue to be managed well. Unfortunately this invites not encouragement but, rather, efforts by the official machinery to disrupt their functioning. Thus it is reported that the *phad* irrigation system of northwestern Maharashtra was deliberately destroyed by the authorities to ensure that peasants could not function in an autonomous, independent manner in this democratic nation.

Community-managed forest lands, which once existed over large parts of India, were steadily dismantled during the nineteenth century. The process of state usurpation was consolidated in the Indian Forest Act of 1878, under which the colonial government took over massive areas of forest. Over the years the extent of forest appropriated by the state steadily increased until it came to exceed one-fifth of India's total land area (Gadgil and Guha 1992).

As they have been destroyed for over a century, there is little concrete information on pre-colonial community-based forest management systems. Glimpses of how such systems functioned may, however, be obtained in the writings of colonial officials. An officer in the Garhwal Himalaya wrote in the 1920s of how, despite official apathy, customary restrictions on the overuse of forests operated over large areas, with village grazing grounds well maintained and fuel and fodder reserves carefully walled in (see Guha 1989a: 31). A more detailed account can be found in the report by G.F.S. Collins, a British revenue official assigned to review lands available for community use in the Uttara Kannada district of Karnataka. Between 1860 and 1890, these common lands had mostly been taken over as reserved forests, only very limited areas remaining available for community use. But here too, the communities were stripped of all rights of regulating the use of resources on lands retained for their use. These were treated as open-access lands, and anybody could come and harvest produce without local communities being

Plate 9 A fuelwood depot in Karnataka. Exhaustion of open-access forests in the neighbourhood of villages is forcing the government to arrange for supplies of fuel from more distant forest areas

in a position to control or monitor use. This resulted in the overuse of these tracts of land, which was used in turn to justify the progressive conversion of community lands into reserved forests. Thus the extent of such lands was reduced from 7,185.9 km^2 to 353.3 km^2 between 1890 and 1920.

The resulting protests led to a reassessment by Collins in 1920 (Gadgil and Iyer 1989; Gadgil and Subash Chandran 1988; Collins 1921). Collins reported that he found three villages in the coastal Kumta district which had on their own initiative continued systems of community-based management, although such systems had no official backing. The inhabitants of Chitragi, Kallabbe and Halakar had been strictly regulating harvests from forest lands available for community use adjoining their villages, paying for a watchman themselves. As a result these forests had an excellent standing biomass, apparently being used on a sustainable basis. Collins commended these systems and suggested that they be formally recognized through promulgation of a village forest act, so that the system could be implemented elsewhere as well. Such an act was passed in 1926, and the village forest councils given formal authority.

The management of these three village forest councils of Chitragi, Kallabbe and Halakar continued in excellent shape until the 1960s. At that time the provincial boundaries were redrawn and Uttara Kannada was included in the newly formed state of Karnataka. Promptly the Karnataka Forest Department served notices disbanding these village forest councils. By that time the town of Kumta had grown right up to the village of Chitragi, whose inhabitants accepted the dissolution of the forest council. This was followed by a quick spurt of felling and total destruction of that forest tract over a period of three months, the state authorities clearly being in no position to arrest this destruction. The people of Kallabbe and Halakar, on the other hand, went to court, challenging the dissolution of village forest councils, and continued to regulate harvests. In the mid-1970s, while the court case was still pending, a timber contractor for a plywood mill began to extract wood from the Kallabbe forest. When the secretary of the village forest council asked the help of the Forest Department in preventing these commercial harvests, the authorities instead backed the timber contractor. Thus have arms of the iron triangle, in this case the Forest Department, the plywood mill and the timber contractors, continued to work with each other, and against the interests of ecosystem people, sustainable resource use and the spirit of democracy (Gadgil and Iyer 1989; Gadgil *et al.* 1990).

Following independence, there have been sporadic attempts to decentralize political authority, to create democratic institutions at the level of *mandals* (= village clusters), *taluks* (= counties) and *zillas* (= districts). Traditionally a council of five or more village elders, the *panchayat*, used to manage community affairs. In consequence, village-level elected committees are called *mandal panchayats*, and the district-level committees *zilla parishats* (*parishat* = assembly, council). All over the country such institutions have been set up but almost always they have been undermined by an alliance of state and national-level politicians and government officials. This is because the politicians operating at higher levels can afford to ignore the concerns of poorer people at the grassroots, instead working in league with the bureaucracy to pursue the interests of the omnivores. But politicians at the *mandal* and *zilla* level have perforce to take account more directly of the interests of ecosystem people, which are often in conflict with those of omnivores. Hence decentralized political institutions have been systematically sabotaged by the omnivores. Even where they have been set up, they are subject to arbitrary dissolution and indefinite postponement of elections.

There are, however, forces, especially outside the dominant Congress Party, that have from time to time tried to take advantage of the potential support from the masses of people and have pushed for such decentralized institutions. Thus the Janata Dal government that came to power in the state of Karnataka in 1983 made a serious attempt to empower and establish *mandal panchayats* and *zilla parishats* in the state. On returning to power in 1989, Congress moved quickly to suspend the workings of the *panchayats.* However, in the period during which these institutions were functioning, local environmental issues had indeed become much more visible in the political process. Thus the Uttara Kannada *zilla parishat* passed a unanimous resolution opposing the setting up of an atomic power plant in its district, and the Tumkur *zilla parishat* supported the resolution by one of its constituent *mandal panchayats* that it did not want a cement factory to be built in its locality. In both cases the state and central authorities overruled the wishes of the local people.

The only state in which decentralized political institutions have been nurtured over a sufficiently long period is the state of West Bengal. It also happens to be the state with the longest period of stable rule by any coalition of political parties, having been governed at the time of writing for over seventeen years by an alliance of left parties. The leftists who initially came to power with the support of industrial labour and the urban middle class found their hold over this constituency of the omnivores to be rather tenuous. They then turned to building up a rural support base by pushing through agrarian reform (most notably, the recording and enforcing of the rights of sharecroppers) and creating genuinely powerful political institutions at the village and district levels. With the support generated by rural reform the left could capture control of the decentralized political institutions through their village-level political cadre. Once this system has been entrenched the ruling left parties have a strong interest in its maintenance (Kohli 1987).

This system is of course by no means free of corruption. But it is a system in which the interests of the ecosystem people have a significant measure of influence. Only in the state of West Bengal must officials respect and listen to members of the *mandal panchayat.* It is therefore not a coincidence that the involvement of people in management of forest resources has made greater progress in West Bengal than anywhere else in the country (Malhotra and Poffenberger 1989).

India is a federation of states with considerable powers devolving to the state governments. Over the past forty years there has, however, been a growing tendency towards centralization of power, progressively weakening state-level political institutions. There is a widespread belief that this trend has favoured the cause of environment and that state governments are far less likely to promote environmental degradation. But there is little solid evidence to support this view. It is true that in 1982–3, when Mrs Indira Gandhi was Prime Minister, she took the side of environmentalists and against the wishes of the Kerala state legislature halted the construction of the Silent Valley hydroelectric project, which would have destroyed some of the finest rain forest of the Western Ghats. But on a different occasion Mrs Gandhi also bowed to political pressure to permit encroachment over large tracts of rain forest by influential plantation owners in the same state of Kerala. Again, while the Uttar Pradesh Chief Minister, Mr Kalyan Singh, called for a suspension of the controversial Tehri Dam in the Garhwal Himalaya after a massive earthquake in October 1991, the central government refused to accept this suggestion. On the whole, it would appear that both state and central governments tend to be dominated by the same omnivore interests, and little will be gained or lost for the cause of environment if the balance shifts one way or another. On the other hand (as we shall demonstrate in Part II of this book), a great deal is likely to be gained if political power is actually transferred to the masses of ecosystem people through a system of decentralized political institutions fully operative at the village and district levels.

Those who maintain that a strong central government is in congruence with the interests of a healthy environment might also have to reckon with the experience of the Indian Emergency of 1975–7, when democratic processes were in abeyance. It was during this period that the Kudremukh Iron Ore project was initiated in Karnataka as a public-sector undertaking without the state government even having been consulted. This project uses ore poorer in iron content than many others available in Karnataka. The ore is then enriched by grinding it to the consistency of talcum powder, increasing the iron content and converting it into pellets that are marketed. This is an expensive process which has ensured that the company ran losses for more than a decade while other private mining concerns were all making handsome profits. The mines are located in heavy-rainfall areas, much of them covered by a pristine rain forest, from where originates the Bhadra, an important river of peninsular India. With huge amounts of waste of the consistency of talcum powder, the project has caused serious siltation of the Bhadra river and the reservoirs, and pollution of the sea near the site of the pelletization plant on the west coast. The mines have also opened up access to a rich rain forest tract and rendered possible its destruction. It is perhaps not an exaggeration to say

Plate 10 Prior to initiation of extraction of the ore, natural grassy banks and evergreen forest graced the site of the Kudremukh mines in Karnataka's Western Ghats

Plate 11 The Kudremukh iron ore mines have laid to waste one of the most picturesque parts of the hills of the Western Ghats

that the project is at once an economic and an environmental blunder of a major magnitude, a blunder perpetrated during the Emergency when public debate of such issues was almost totally suppressed.

The Emergency period also witnessed a fierce coercive drive for population control. This program of forced sterilization caused great resentment and was indeed a major factor in the defeat of the Congress Party when elections were finally held in March 1977. This coercive drive was clearly a major disservice to the goal of halting the growth of India's population, since no politician is willing any longer vigorously to pursue a policy of population control, even by democratic means. Thus there is little to suggest that the suspension of democratic processes during the Emergency made any positive contribution to the cause of the Indian environment. Neither is there evidence from other countries, be it Myanmar or the erstwhile Soviet Union, that dictatorships are anything but abusive of environmental interests.

THE ECOSYSTEM AND ITS PEOPLE PAY

India has thus become effectively organized as a democracy of the omnivores, for the omnivores, by the omnivores. This is a system in which the interests of the huge numbers of ecosystem people and ecological refugees can be largely ignored. The omnivores can capture resources by using the state apparatus, while passing the costs of resource capture on to the rest of the population. This permits the system to tolerate massive environmental degradation and to use resources in an exceedingly inefficient manner.

The trends in the utilization of bamboo in India provide a graphic illustration of this process. Among the fastest-growing plants in the world, bamboos, with their long fibres and ease of working, find an enormous number of applications throughout the tropical world. Bamboo slats go to make walls, doors, windows of huts and large pillars and beams. Grains are stored in bamboo baskets and bamboo culms are fashioned to form cylindrical vessels and spoons. Bamboos are joined end to end to form irrigation pipes and used in seed drills. Bamboo shoots are a nutritious food and bamboo seeds as good as rice.

But for the early British forest managers bamboo was a pernicious weed. No matter that millions of India's ecosystem people extensively used it, no matter that hundreds of thousands of families of basket weavers depended on bamboos to earn a living. For the British wanted tropical forest lands to produce teak and little else. When forests were cleared to raise teak plantations, bamboo thrived, overtopping teak and reducing its growth. So for many years forest managers prescribed the eradication of bamboo from the forests of India. This policy began to change only when research in the 1920s showed bamboo, with its long fibres and soft wood, to be a very desirable raw material for paper making. But for over half a century after bamboo became a commodity of industrial use, it was still viewed as an inexhaustible resource.

It was made available to paper mills at a throwaway price, as low as Rs. 1.50 per tonne to the West Coast Paper Mill in the Uttara Kannada district of Karnataka in 1958. At that time, the market price of bamboo in the cities of Karnataka was of the order of Rs 3,000 per tonne. While poor basket weavers paid a hefty price for this vital raw material, industry was given virtually free access to stocks on reserved forest lands (Gadgil and Prasad 1978).

Economic theory tells us that entrepreneurs will always be interested in maximizing the returns on their investments. At any time there are many alternative avenues by which to accomplish this, and technology is opening up new avenues all the time. Modern enterprises are therefore not necessarily interested in long-term sustainability of a renewable resource, be it bamboo or fish stocks. Indeed, it is rational for them to use up the resource at a rapid rate and to switch to a substitute at a later date if the current profit margin is high enough. Economists therefore suggest that the state should step in to levy a tax and to bring down the profit margins of the use of renewable resources in danger of overuse, so that the users are motivated to use a renewable resource at a rate moderate enough to ensure sustainability (Dasgupta 1982).

But sustainable use of bamboo or fish stocks or ground water is truly important only for the ecosystem people who have no other options – options of moving elsewhere or substituting for a particular resource. Sustainability is of little interest to omnivores who might have plenty of such options. A state acting in the interest of ecosystem people may then be expected to price renewable resources so as to moderate the pace of their commercial exploitation. On the other hand, a state acting in league with commercial interests would subsidize renewable-resource supplies to omnivores, thereby increasing the profit margins and levels of availability to the privileged, while at the same time promoting overexploitation and exhaustion at the cost of the underprivileged.

Indeed, the Indian state apparatus has consistently underpriced renewable resources, or in other ways enhanced the profits realizable from their exploitation. To revert to the example of paper mills and bamboo, the West Coast Paper Mill (WCPM) was not only awarded bamboo at nominal prices, but also granted highly subsidized access to land and water. In theory the WCPM was expected to meet its bamboo resource needs in perpetuity from the district of Uttara Kannada. But this expectation was based on faulty, probably deliberately inflated, figures of resource stocks and a very incomplete understanding of bamboo ecology.

Furthermore, the paper mill contractors had little interest in carefully harvesting bamboo, a few culms at a time from each clump, so as to keep the clumps growing. Rather, they were interested in minimizing harvest costs, and tended to clear-cut the most accessible clumps. The result was that the WCPM very soon exhausted the bamboo resources of Uttara Kannada and then turned to the neighbouring state of Andhra Pradesh. As the bamboo resources of that state became depleted, the WCPM moved further afield to the states of Orissa, Assam and Arunachal Pradesh. As bamboo diminished all over, the mill began to use other softer woods such as *Kydia calycina* and eucalyptus. With a captive market ensured through strict restriction of imports the paper industry could hike up prices and maintain large profits despite the substitution of bamboo by more expensive raw material. While the paper industry has been doing very well, and undoubtedly kicking back part of its profit to politicians and bureaucrats who have favoured it, bamboo exhaustion has hit the rural people very badly. For them the quality of housing has significantly declined, and a valued food resource has all but vanished. More critically, for the basket weavers incomes have drastically diminished, turning many into ecological refugees (Gadgil and Prasad 1978).

This ability to pass on the costs of profligate resource use to the masses also means that there is little concern with obtaining adequate returns on the resources deployed. That is why there is scant concern within the iron triangle with dams that have silted up rapidly and with irrigated lands that have become waterlogged, with high levels of transmission losses in the supply of electricity, low efficiencies of boilers in factories and massive thefts of coal transported on railway wagons. For the users of water, power, coal and transport are not paying the full cost of services, and do not care if the real costs are unduly large. As for those in charge of delivering these services, the greater the tolerance of inefficient resource use, the greater the possibilities of their misappropriating public funds.

DECOUPLING RESOURCES SPENT FROM SERVICES DELIVERED

So for the state apparatus, which had until very recently taken on the task of delivering everything from water, coal and electricity to classical music and technological innovations, what becomes important is not what the people get, but what the government spends. The state also has had the powers to regulate everything from cutting of trees and building of houses to marketing of liquor and manufacture of soap. Again, given that no influential segment of society really cares for what happens to the trees or who drinks the hooch, the apparatus has developed a vested interest in accumulating regulatory powers to harass and extort, rather than genuinely protect trees or maintain the quality of soap. As a distinguished scholar and civil servant has written, the implementation of development programmes in India is in the hands of 'an unholy alliance of the elite, the educated and the bureaucrat, incapable of eschewing coercion, corruption and fraud' (Mitra 1979). This alliance has then concentrated on accumulating regulatory powers, and spending a great deal of money while delivering very little. Indeed, the whole functioning of the government in independent India seems to have evolved so as to decouple resources expended from services delivered!

The Republic of India has been set up as a confederation of states; instead, the central and state governments behave as confederations of numerous ministries, departments and public-sector corporations. The efficiency of the overall development process obviously depends on a proper co-ordination of the activities of the different agencies. Thus the provision of irrigation must be related to soil conditions, and attempts at forest regeneration to the provision of fodder to livestock. Soil and water conservation activities in a watershed must treat land controlled by revenue, irrigation, forest and public works departments as well as by private holders in a co-ordinated fashion. But despite the existence of bodies like the district development councils, not a shred of such co-ordination exists. Instead, each department has developed a culture of a well-knit, highly organized group pursuing its own vested interests in an independent fashion. Of course, each department does interact with others to carve out the total share of the pie, but to no other useful purpose. So district development councils or state or central-level planning commissions are reduced to the function of accounting for the allocation of funds among different government agencies, and play little role in genuinely integrating the process of development.

There are innumerable examples in our experience of how the lack of such co-ordination leads to considerable resource wastage. All over India, surface irrigation has been developed without reference to soil conditions. So water flows along thousands of kilometres of unlined canals even where the soils are heavy clays such as the black cotton soils. These soils have poor drainage and become waterlogged very readily. With time, salts diffuse to the surface of waterlogged soils leading to salination (N.J. Singh 1992). Since none of this has been taken into account in the design of irrigation systems, despite the existence of government departments like the Soil Survey and Land Use Bureau, huge tracts of lands on either side of irrigation canals have become saline, waterlogged wastelands in many parts of India.

Mahatma Gandhi once called the goat a poor man's cow. Goats are hardy animals capable of grazing on all manners of plants that sheep or cattle or buffaloes will not put a tongue to. Goats can earn supplementary income for the otherwise assetless rural poor, and their numbers have been rapidly increasing with the exploding demand for goat meat from omnivores. The provision of subsidies for the purchase of goats is therefore something that can be readily justified as a part of rural development programmes, especially to help members of tribes and lower castes. Since government agencies love disbursing subsidies, goat subsidies are a popular component of programmes of the Animal Husbandry Departments, including those operating on the Western Ghats hill chain. These have been a regular component of the Western Ghats Development Programme, a special programme of assistance from the central government. But much of the grazing on the Western Ghats is on forest land, and this increased goat population is apt to increase the grazing pressure on forests. There has been no systematic investigation of the effects of grazing on forest regeneration or growth, as is indeed the case for most issues pertaining to forest ecology. However, just as government agencies love disbursing subsidies, they love powers to regulate people's activities. Thus the Forest Departments have been urging a ban on the maintenance of goats in areas with large forests, including all *taluks* (counties) falling under the Western Ghats Development Programme. In this manner, while one wing of the government has been disbursing subsidies to encourage the purchase of goats, another wing has been demanding that goats be banned.

Over the years there has been an increasing realization of the resource wastages that result from this lack of co-ordination between sectors. Here watersheds are evidently an appropriate unit for integrating land-based development activities such as agriculture, animal husbandry, fisheries and forestry. As a consequence, integrated watershed development programmes are coming into vogue in many parts of the country, the state of Karnataka having taken a lead in this matter. But the entrenched government departments are totally unwilling to work with each other in a co-ordinated fashion for this purpose. So the Karnataka government has created yet one more agency, the Watershed Development Board, to work in an integrated fashion in a few specially demarcated localities. There can be no more eloquent testimony to the complete unwillingness of the government machinery to work in a co-ordinated fashion, and to its interest in the proliferation of state agencies and departments (State Watershed Development Cell 1989).

The states of pre-colonial India, as well as the state in British India, invested little except in maintenance of law and order and the collection of taxes. There were, of course, some investments in roads, communications and irrigation, but the management of natural resources depended greatly on private and community effort. (The Indian farmer adapted many new crops brought in from outside by Europeans prior to conquest, such as chilli, potato, tomato, guava and cashew. The spread of chillies was by far the most remarkable; chilli had become a common ingredient of Indian cooking within half a century of its introduction in the early 1500s (Achaya 1993).) Village communities put up and maintained small-scale irrigation works with their own effort, and took care of health services, including vaccination against smallpox. After independence all such activities became a function of the state apparatus, calling for an investment of public funds. New crops and crop varieties are promoted by the state, and since the state cannot desilt irrigation tanks they have fallen into disuse. All development programmes have thus become utterly dependent on the availability of state funds.

Since there is little internal pressure to use this money effectively the state machinery has relentlessly pushed up the costs of delivering all services. A foremost component of this is civil construction: that is, the building of houses and offices, roads, bridges and dams. In charge of such activities is the PWD – the Public Works Department, also informally known as Public Waste

Department or Plunder Without Danger. Costs for construction supervised by the PWD are substantially higher than what private effort would require. Anecdotal evidence suggests that this is because around 30 per cent of the costs are misappropriated by the state machinery, shared among the concerned politicians and bureaucrats. In such a system the PWD has strong vested interests in inflating costs, and the departmental norms are indeed designed more to push up costs than to enhance quality.

As a consequence, the PWD consistently opposes all attempts at innovation that would cut down costs. One such noteworthy attempt was by Laurie Baker, an architect from Kerala. Baker has designed a number of elegant private and public buildings at costs well below PWD norms. In the early 1970s, he built the campus of the Centre for Development Studies in Thiruvananthapuram at costs well below those initially estimated. With the money thus saved, the centre was able to build one of the finest social science research libraries in Asia. At that time, Mr Achutha Menon, a progressive politician, was the Chief Minister of Kerala and thought it would be an excellent idea to use Baker's designs and techniques to economically construct a large number of primary school buildings. He called together a group of educationists and engineers along with Mr Baker to discuss this scheme. The story goes that Baker was virulently attacked by the engineers, while the others were convinced that his ideas were worthy of implementation. In the end the Chief Minister gave up the idea of using Baker's designs, apparently afraid that the government machinery would somehow sabotage the construction, which might even end up in the collapse of some school buildings.

Not that structures constructed at high costs following PWD norms and built under its supervision do not collapse. Several have done so, including the big bridge over the river Mandovi in Goa. Goa was under Portuguese control until 1961 and its citizens tell the story of how an engineer working in the Goa government was at that time sent to jail for several years when a building constructed under his supervision collapsed, leading to the death of four people. But there is no such accountability in independent India, and the collapse of the Mandovi Bridge with nobody held responsible is an unfortunate symbol of the nature of India's development efforts, which have erected a high-cost, low-quality economy plagued by environmental degradation and social inequity.

Given the extreme levels of inefficiency with which development projects are executed, it is not easy to rationalize their being taken up. To get over this problem, project planners tend to greatly overestimate likely benefits. Thus in the late 1950s foresters advocated giving up selection fellings in stands of natural forest, replacing them instead by aggressive plantation forestry, which called for large-scale clear-felling of natural forests to create plantations of fast-growing exotics such as eucalyptus and tropical pine. This approach was adopted in the absence of any careful trials of how successful such human-made plantations were likely to be. Without proper evidence it was estimated that the eucalyptus plantation yields would be between 14 and 28 t/ha/year. On this basis, large tracts of prime rain forest, even primeval sacred groves, of the Western Ghats were cut down, only for it to be discovered that in these high-rainfall areas eucalyptus falls prey to a fungal disease agent and may either die off or grow very slowly. The realized productivities of the eucalyptus plantations so raised were only between 1.5 and 3 t/ha – barely 10 per cent of the projected ones, and well below the productivities of the natural forest so replaced (Prasad 1984).

Many irrigation projects are similarly justified on the basis of greatly exaggerated estimates of the likely enhancement of agricultural productivity. It tends to be assumed, for example, that the farmers will receive reliable water supply, and will follow a recommended pattern of cropping involving use of moderate amounts of water over extensive areas. Neither of these assumptions ever holds in practice. Farmers in the upper reaches end up cultivating water-intensive crops such as sugarcane and rice and using too much water, often leading in the long run to problems of waterlogging and salination, while farmers in the lower reaches receive very inadequate supplies. The net result is a far lower enhancement of agricultural productivity, well below what has been projected. Nevertheless, despite the experience that the kind of cropping pattern prescribed can never be implemented, estimated benefits continue to be inflated, being computed on an imaginary basis.

Benefits of irrigation projects are also overestimated by underestimating rates of siltation and the life of the reservoirs. There has been very little careful research on the rates of erosion and siltation of streams under different patterns of land use in India, and the rates used at the project formulation stage are invariably arbitrarily assumed minimal rates. Several years ago one study by India's Planning Commission revealed that the realized rates were on average 2.17 times greater than the rates assumed at the time of project formulation. Among the worst cases of such an underestimation is the Tungabhadra project in Karnataka. The estimated life of this project, completed in 1953, was 100 years, that is, its dead storage capacity was expected to be fully silted up over such a period. However, by 1983 it was estimated that its live storage had already been reduced to 90 per cent after full exhaustion of its dead storage capacity through siltation.

Several features of the functioning of the state apparatus are consistent with this absence of any concern over resource wastage; indeed, they convey the distinct impression that resource wastage is welcome. These include the extremely irregular releases of funds, the chasing of paper targets and the frequent transfer of government officials. The government financial year runs from 1 April to 31 March, and typically, large amounts of funds are suddenly released towards the end of the financial year, even as late as 29 March, with the expectation that the money will be effectively utilized by 31 March. This is quite clearly impossible so that huge amounts of money are regularly wasted and misappropriated. Indeed, in official folklore the month of March is known as *Loot-mar-ka-mahina* – the month of plunder and robbery.

The gross inefficiency of this system of the sporadic use of funds is brought out by a study of the construction of fuel-efficient wood stoves in the state of Karnataka. The design being propagated through this programme in the period 1984–7 was known as Astraole, and was developed by engineers at the Indian Institute of Science in Bangalore. The fuel efficiency of the village wood stoves in Karnataka is between 12 and 17 per cent; under laboratory conditions the Astraole can generate efficiencies of the order of 42 per cent. In the field the efficiencies are lower, of the order of 32 per cent, since a variety of dishes are cooked and they cannot be so adjusted as to fully use up all the heat generated. The doubling of fuel use efficiency is still a significant enough gain. But its realization depends, while the stove is being constructed, on carefully maintaining the sizes and relative dispositions of the different parts. Careless construction can disrupt the proper flow of fresh and hot air and drastically bring down efficiency gains. There may even be situations where the new stove is worse than old ones if improper construction leads to reverse airflows from under the pots towards the fuel box.

Recent years have witnessed an attempt to propagate non-conventional energy sources such as solar energy and devices enhancing efficiency of energy use such as the Astraole. These drives have taken on the standard form of stimulation through free or subsidized supplies. Once subsidies come in, the government machinery develops a vested interest in pushing through the programme while misappropriating a part of the subsidy. The drive is also fuelled by the setting of targets to be fulfilled within some time limit – such as 10,000 Astraoles to be constructed before 31 March. All of this conspires to promote shoddy execution, which may in fact lead not to better, but worse, performance.

The most serious lacuna in the system is of course the near-total absence of any evaluation of the performance, especially by those who are supposed to benefit from the services generated. Thus villagers who end up using Astraoles in their huts have no authority to evaluate how useful their introduction has been. By and large nobody else evaluates the performance either. However, in the case of the Astraole programme in Karnataka the State Council for Science and Technology did conduct a post-programme evaluation, including interviews with users. Figure 2 sums up the results of these interviews, revealing that in a substantial number of cases the stoves turned in very poor performances. Figure 3 brings out one of the factors behind this, namely the hurried construction of a large number of stoves towards the end of the financial year (Ravindranath *et al.* 1989).

Another serious victim of *targetbaji* – the chase of paper targets at the cost of performance – is India's family planning programme. Individuals are given financial incentives to undergo vasectomy or tubectomy operations, and

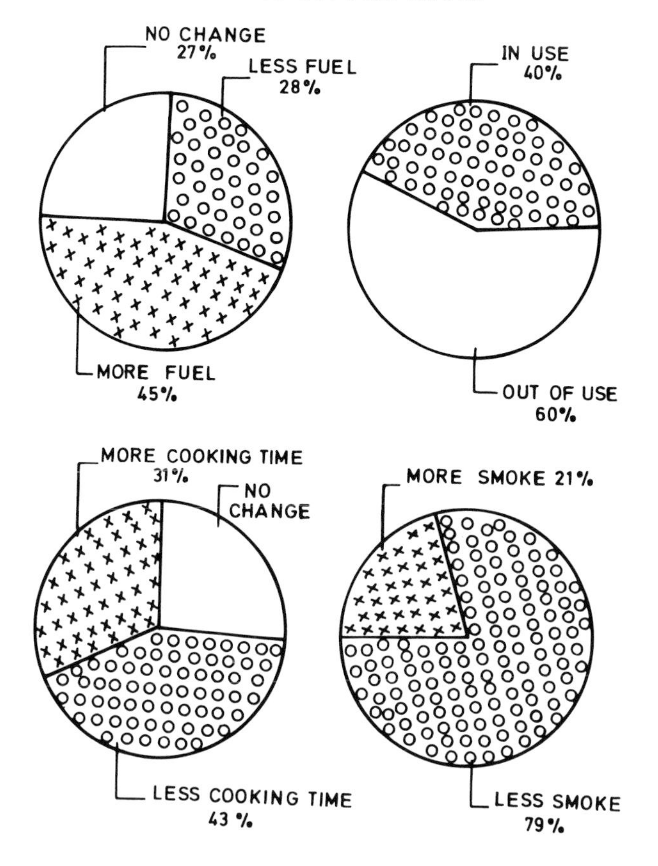

Figure 2 Results of a study for evaluating the performance of a new fuel-efficient stove design, Astraole, based on a survey of 280 stoves in fourteen districts of Karnataka state in 1988 (after Ravindranath *et al.* 1989)

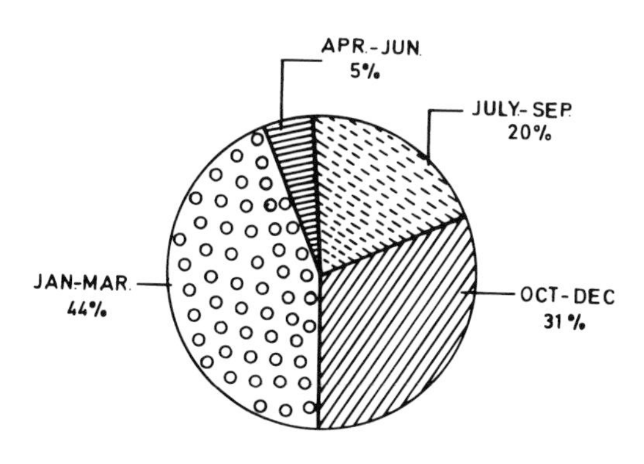

Figure 3 Time of year of construction of a total of about 153,000 Astraole stoves over the three-year period 1984–7 in the state of Karnataka (after Ravindranath *et al.* 1989)

health centres are assigned targets of so many thousands of operations to be performed per year. A substantial proportion of India's masses are not really motivated to restrict the number of their offspring, who are seen as useful devices to add to the very low family incomes. Meeting targets thus necessarily ends in involving a measure of coercion. On top of this, the lack of performance evaluation and compulsions to meet targets promote careless surgical operations. Shoddy construction of bridges or wood stoves is bad enough but shoddy operations can be a hazard to health and even cause deaths, which can readily turn people against these programmes.

GETTING AWAY SCOT-FREE

All over India people are of course quite aware of the thoroughly unsatisfactory performance of the state apparatus and the high levels of misappropriation by the iron triangle of politicians, bureaucrats and contractors working in league with each other. The failure to bring the situation under check relates to three factors: lack of evaluation of performance; the difficulty of assigning responsibility for poor performance; and the lack of appropriate machinery to reward good or punish poor performance. We will come back later to the vital issues of lack of performance evaluation and suitable incentives, positive and negative, to promote good performance. Equally significant in encouraging wasteful resource use is the difficulty of assigning responsibility. A main reason for this is the system of frequent and arbitrary transfers. The British introduced this system because they did not want officials to strike roots and develop sympathies with the subject population. The state in colonial times also performed the relatively simple task of maintenance of law and order and collection of revenue – tasks that did not require very much specialized locality- or society-specific knowledge. But the state machinery is now involved in a variety of interventions in complex local ecosystems and social systems, whether it be introducing silkworms, goats, groundnut or coconut, providing drinking water, eradicating guineaworms, eliminating sharecropping or freeing bonded labourers. Effective action therefore calls for a deep understanding of and empathy for local people and their environment. But frequent transfers make this understanding impossible to acquire. Indeed, by the time an official begins to appreciate the local situation he or she is invariably packed off. Thus government programmes become a shifting kaleidoscope reflecting the whims and fancies of a multitude of poorly informed even if sometimes well-intentioned administrators.

Unfortunately, a significant proportion of the government machinery is merely self-serving. The administrators are not responsible to the people they deal with, but only to their political bosses. The political masters have perfected the art of manipulating elections with the help of catchy slogans, communal politics, money power and hooliganism. They are themselves hardly responsible to anybody. Politics in India has thereby become the pursuit of public affairs for private profit. Yet it is notable that no state or national politician has ever been convicted and sent to prison for corruption, although a large number (perhaps even a majority) are certainly guilty on this count. Meanwhile, administrators are not rewarded for being honest, capable, or for doing the job they are assigned to do. Instead they are rewarded for conniving with politicians in extending patronage to people in favour and in misappropriating public resources. In this system transfers are used by politicians to post pliable administrators to where they want them to be, and to remove administrators who will not do their bidding.

Apart from organizing specific services such as delivering water or vocational training, the bureaucracy is also responsible for regulating resource use and abuse, be it the cutting of trees or the pollution of streams. As with delivering services, the self-serving political-bureaucratic apparatus can convert such regulation into an opportunity to further private ends, rather than to render public service. In popular parlance this is known as earning *aankhbandi* allowance – an allowance for looking the other way when a law or regulation is being violated. Thus, just south of the centre of the Western Ghats range lie the Nilgiris or Blue Mountains, a massif rising to an altitude of over 2,400 m. In recent times a Private Trees Protection Act (PTPA) has been promulgated, applying to the Nilgiris district in the state of Tamil Nadu. This act prohibits the cutting of any trees without government permission, even on private lands.

It happens that the Nilgiris are full of tea estates, estates with tens of thousands of silver oak, *Erythrina* and other shade trees. The landowners regularly harvest part of the crop, replacing it with new saplings. With the PTPA they have to run from pillar to post and grease palms to get the required permission. At least, so allege all landowners – from tribals to big estate owners. They say that despite the difficulties they face, the act has in no way reduced the rate of tree cutting. It is just that individual landowners are now incapable of handling the transactions involved. They have to go to timber merchants who arrange to get the necessary permits and of course swallow a good portion of the price, part of which is undoubtedly shared with the authorities. In this manner regulation results not in gains for the environment, but for the politician–bureaucrat–contractor iron triangle.

Pollution control too has apparently degenerated into a similar system of ill-gotten gains for the regulators. Harihar Polyfibres, whose factory discharges effluents into the River Tungabhadra, was embroiled in a long dispute with local people and voluntary agencies who strongly contended that the State Pollution Control Board behaved as if it was an arm of the industry (Hiremath 1987).

THE MONOPOLY OF PRODUCTION, DISTRIBUTION, INFORMATION

Such an inefficient system can function only if it does not have to compete with other enterprises, Indian or foreign, delivering the same goods and services. Nor can it survive if there are alternative systems for accomplishing the same regulatory objectives. It is therefore only natural that the state apparatus had until very recently assumed monopolistic control over its various functions. Thus government electricity boards and power corporations have had monopolies over the production and distribution of electrical power over most of the country. These systems are notoriously inefficient, with constant breakdowns of power stations, huge transmission losses and rampant thefts of power. Moreover, until very recently they have focused on large-scale, centralized power generation and its distribution over countrywide grids. Such a system cannot effectively deliver power to small, remote hamlets, especially in regions like the Himalaya. Small-scale power generation for local use using hill streams can therefore be a superior alternative in such areas. Indeed, dwellers in such Himalayan villages have traditionally employed water power to grind wheat.

There have been some attempts to use these Himalayan hill streams to generate power on a small-scale, local basis. Not far from the town of Gopeshwar in the Alakananada valley, where many of the activists of the Chipko *andolan* are located, is the village of Tangsa. In Tangsa the villagers have set up a small-scale sawmill using water power to saw timber for local construction purposes. They have been interested in employing the same technology to generate electric power for lighting village streets and houses. But despite many attempts by the Chipko activists, the Uttar Pradesh electricity board stands in the way, preventing any violation of its monopoly over the generation and distribution of power.

The state similarly claims monopolistic rights over all streams, rivers, estuaries and sea waters. Only state agencies such as irrigation departments have therefore the right to dam and divert watercourses, or to auction sand from stream beds, and so on. There are a series of problems with such a management regime. First, decisions such as damming, or auctioning the sand tend to be taken without any reference to the aspirations or preferences of the people involved. Thus those whose lands are to be submerged or acquired for a canal are not advised of this fact even long after the decision has been taken, and may even be forcibly evicted at the eleventh hour without any proper arrangements for rehabilitation or compensation. Nor are the recipients of irrigation water consulted as to the course of distributaries, the amount of water they desire, or the cropping pattern they plan to adopt. The beneficiaries of irrigation by no means pay for the state investment in bringing the water to their fields, investment that may be as high as Rs 52,000 per ha. But by and large they welcome the irrigation water and try to grab as much

Plate 12 Unregulated removal of sand from river beds has led to serious disruption of water regimes in many parts of the country

of it as possible. Those in the upper reaches of command areas go in for cultivating water-hungry crops such as paddy and sugarcane; those in lower reaches often do not receive the water they were expected to get. The resultant chaotic pattern leads to very inefficient resource use, which worsens water scarcity. At the same time, since the beneficiaries do not pay for the water that reaches them, nobody questions the heavy costs of delivering this water. The project implementers take advantage of the situation to hike up the project costs to exorbitant levels.

It is only now that this process of water misuse is beginning to be questioned. Among the pioneers of such attempts are the farmers of Tandulwadi in the Sangli distict of Maharashtra. In the 1980s these peasants ran into serious problems of water shortages. Coming into contact with a voluntary organization called Mukti Sangharsh, they began to search for the causes, only to discover that the rainfall records did not indicate any trend of decline. They then realized that scarcities were primarily due to the new pattern of cropping involving heavy water demand.

At the same time the farmers realized that the flow of streams through their village was being adversely affected by the indiscriminate removal of sand by private contractors. The farmers decided to take matters into their own hands – to build a dam on their own, using the money raised by selling sand removed to promote dam construction; to share the irrigation water from the dam on an equitable basis, even allotting a share to the landless families; and finally to use the water effectively through a cropping pattern that kept out crops with a heavy demand for water. For the builders of this dam, called Baliraja by the peasants, this involved challenging the state monopoly on more than one count, for in theory they have no rights over the sand or water, except through the mediation of the state machinery. But they have succeeded in challenging this monopoly and the Baliraja dam stands today bringing water to parched fields. Its achievement brought forth this disgruntled, and most revealing, remark by a minister in the Maharashtra government: 'What will the government have left to do if peasants go on building dams themselves?' (Singh 1994: 221).

Perhaps the most pernicious of all state monopolies is that over information. In the traditional caste society of India, based on a hereditary division of labour, information was compartmentalized in the extreme. The upper castes, especially *Brahmans*, had monopoly over formal knowledge, be it of astronomy, the Ayurveda system of medicine, or religious texts, chants and rituals. The many separate artisanal castes – potters, goldsmiths, weavers – each had exclusive knowledge of materials and techniques pertaining to their specialities. Peasants had knowledge of farming, herders had knowledge of sheep or camel rearing, specialized nomadic castes had knowledge of herbal medicine, and so on. Information was passed on from generation to generation and almost exclusively within kin groups, though occasionally through specialized teaching institutions such as *pathashalas* for training priests in religious texts and rituals. The monopoly was maintained by this tradition of the transmission of knowledge being limited to members of kin groups. But there was coercion as well to ensure monopolistic control. Thus Manusmriti, the Hindu text of the second century of the Christian era, which prescribes codes of behaviour, specifies that no lower caste or *Sudra* should even hear the sacred Vedas of the *Brahmans*. If a *Sudra* commits such an offence he could be punished by having molten lead poured into his ear. There is also the famous story in the Mahabharatha epic of Ekalavya, of a tribal youth who had acquired the knowledge of archery supposed to be limited to high-caste warriors or *Kshatriyas*, and who, once discovered, was forced to lose his right thumb, so that the monopoly would not be broken.

This fragmentation of knowledge, so that the *Sudras* – artisans, peasants and herders – with knowledge of the relevant techniques in their limited domains were prevented from access to the knowledge and technique of other caste groups, as well as from access to the formalized body of knowledge with the priests, was undoubtedly a significant reason why Indian society failed to take a lead in the development of science and technology, despite considerable advances on many fronts. When the British conquered India, they were helped in good measure by their mastery over the newly emerging scientific knowledge and its accompanying techniques. They arrogated to themselves the status of a superior isolated group, like the *Brahmans* of yore, and attempted to maintain a monopoly over science and technology. The British, however, needed to interact with the Indian population in order to maintain law and order, collect taxes and organize a flow of raw materials to Britain and of manufactured articles to India. For this purpose they required allies, who came almost exclusively from the upper castes; that is, from among the omnivores of pre-colonial times. These were provided with an education appropriate to serve their alloted functions, which had little scientific or technical content.

The function of the colonial establishment and its Indian allies was predominantly one of subjugation. It did not want to be questioned by the subjects, and therefore shielded itself behind an Official Secrets Act. The government did of course acquire such information about the subject population as it deemed desirable, as witness the proliferation of district gazetteers, land settlement reports and other official records. But the subjects had absolutely no access to information about what the state was up to. The government communicated with the public only through occasional gazette notifications and government orders.

After independence the state apparatus radically changed its function from that of subjugation to that of all-round development. But this all-round development was only in theory; in practice it became a mix of subjugation on behalf of the Indian omnivores, coupled with the disbursement of patronage trickling down to the ecosystem people and the increasing numbers of ecological refugees. This system can also perform best only by being kept out of the public gaze. So, despite the ushering in of a democratic government, and elected legislatures, the state apparatus continues the way of the colonial state in absolutely refusing to share any information with the public. The Official Secrets Act remains intact, all government papers are marked 'Official Use Only', and members of the public have no access to information of vital interest to them – pertaining to submersion areas of a proposed dam, or plans to clear-fell a particular piece of forest, or of who is eligible to receive subsidies for digging a well, or what were the findings of a government committee that looked into an incident of river pollution resulting in massive fish kills. This monopoly of information comes in very handy in favouring the allies of those in power, as well as in misappropriating public resources.

Over almost five decades now, the omnivores have thus been merrily pursuing the capture of India's resource base. Only fragmentary estimates are available of state resources devoted to enhancing resource supplies through subsidies – most of which ultimately go to omnivores – but the magnitude is staggering, perhaps over 15 per cent of the gross national product. This share has been steadily increasing; over the 1980s it grew at 18 per cent per year, well over the inflation rate (Mundle and Rao 1991). At the same time, the size of the state apparatus in terms of numbers employed in government, joint sector and government-aided institutions has been rapidly increasing, so that salaries and perks of employees are swallowing an ever greater proportion of state funds. All of this is financed to a significant extent by deficit financing creating inflationary pressures. Such inflation primarily hurts the masses outside the charmed circle of omnivores. In part the state operations are financed by external loans, so that the debt burden has also been continually going up. There is increasing pressure to exploit the natural resources of the country, be it iron ore, fish, leather goods or textiles, to service this debt and pay for the exorbitant levels of import of oil. The impact of this export of natural resources is again felt primarily by the ecosystem people, whether through overuse, through diversion of resources traditionally used by them or through pollution-related damage (Kurien 1993). It is little wonder, then, that India has become a cauldron of conflicts directly or indirectly triggered by the abuse of natural resources to benefit the narrow elite of omnivores. It is to a consideration of these conflicts that we now turn.

3

A CAULDRON OF CONFLICTS

As the centre of power and patronage, the Indian city of New Delhi is the venue of year-round demonstrations by organizations representing different classes, castes and ethnic groups. Farmers demanding the provision of subsidized power and fertilizer, industrial workers campaigning for higher pay, upper-caste students fighting against job quotas for backward communities, and ethnic minorities fighting for a separate state all recognize the symbolic significance of a show of strength in the national capital. Assured of widespread coverage by the print media, these demonstrations are often held at the Boat Club lawns, a stone's throw both from the Houses of Parliament and from the government secretariat.

The month of May 1990 saw a phenomenon unprecedented even for New Delhi: a demonstration followed, within a week, by a counter-demonstration. First, villagers to be displaced by the massive Sardar Sarovar dam, being built on the Narmada river in central India, assembled in a peaceful yet joyous *dharna* (sit-down strike) on Gol Methi Chowk – in the heart of New Delhi, and very close to the residence of the then Prime Minister, Mr V.P. Singh. Mostly poor peasants and tribals, these ecosystem people of the Narmada valley sat there for several days, singing, dancing and listening to exhortative speeches by their leaders. Most of the demonstrators had come from Madhya Pradesh, the state containing a majority of the villages to be submerged by the dam. They dispersed only after Mr V.P. Singh met a delegation of the protesters, and assured them that the Sardar Sarovar project would be reviewed. Immediately, politicians in Gujarat, the state that stands to benefit most from the project, set about organizing a counter-demonstration on behalf of the omnivores. After a public meeting at the boat club, the Gujarat protesters also went to meet Mr V.P. Singh. The Prime Minister granted them an immediate audience – he had kept the Madhya Pradesh peasants waiting for days – and told them what they wanted most to hear, namely that he and his government were fully committed to the implementation of the Sardar Sarovar project.

A few months later, the two contending groups were involved in a face-to-face encounter hundreds of miles from New Delhi, on the Madhya Pradesh–

Gujarat border. On 25 December 1990, the Narmada Bachao Andolan (Save Narmada Movement), an organization working in the interests of the potential oustees of the dam, began a 250-km march from Rajghat in Madhya Pradesh to Kevadia colony, the site in Gujarat of the Sardar Sarovar dam. The marchers, several thousand in all, were stopped by the Gujarat police at the border village of Ferkuva, and prevented from entering the state. On the Gujarat side, a large group, including students and plain-clothes policemen, had assembled to heckle the marchers. A stalemate lasting several days ensued, with the pro-dam agitators – who were addressed by the then Chief Minister of Gujarat, Chimanbhai Patel, on 29 December – raising slogans in favour of the dam and against the Narmada Bachao Andolan and one of its leaders, the respected septuagenarian social worker Baba Amte. For their part, the protesters insisted on their right to march peacefully to the dam site at Kevadia.

On the second day of the new year, a group of twenty-five protesters, with their hands tied to emphasize the non-violent nature of their struggle, entered Gujarat, to be stopped by the police 150 m inside the state. Two more groups, again with their hands tied, joined them the next day. On 5 January 1991, Baba Amte and another group of twenty-five also entered Gujarat. After being allowed to cross the border but not proceed further, they began an indefinite *dharna* (strike) on the Gordah River bridge, a bare 30 m inside Gujarat. The next day, other anti-dam activists, including Medha Patkar, indisputably the movement's most important leader, went on hunger strike

Plate 13 Medha Patkar leading a *dharna* by Narmada Bachao Andolan supporters and tribal people

on the Madhya Pradesh side of the border. With the Gujarat government unrelenting, the stalemate continued for several weeks. Finally on 28 January Patkar and her associates, with their own lives in danger, were persuaded by other social workers and voluntary organizations to give up their fast (Anon. 1991).

The Narmada controversy is a particularly charged example of a wide spectrum of social conflicts over natural resources in contemporary India. The past two decades have witnessed a rapid sharpening of these conflicts, although of course they have a long history stretching well into the country's past. Notably, these conflicts were muted in the first two decades of independent India, when the state was widely perceived as the authentic legatee of an all-class and genuinely mass-based national upsurge. On independence the state had also changed its character from being an instrument of subjugation and extortion of surplus on behalf of a colonial power, to one extending its patronage to the entire population. With time, however, the Indian state has lost much of its legitimacy, being increasingly seen as an instrument of a narrow elite of omnivores. At the same time, the democratic system has conferred on the growing numbers of ecosystem people and ecological refugees a modicum of political clout. These twin and somewhat contradictory processes, namely, resource capture by the elite and the creation of a space for social protest, have generated numerous conflicts between and among the three broad ecological classes of modern India: the omnivores, the ecosystem people and the ecological refugees.

The conflicts occur on a variety of scales, from encroachment on a plot of grazing land by families of local landless labourers, to widespread protests by thousands of peasants and tribals against displacement by a mega-project like the Narmada dam. They relate to many different natural resources: land for cultivation or grazing livestock; water for domestic, agricultural or industrial use; mineral deposits; fish stocks; woody biomass; or wildlife. Most dramatically they pit haves against have nots: trawler owners against artisanal fisherfolk, tribals against paper mills, peasants against irrigation authorities. But they also pit poor against poor, as village common lands are ruined in a scramble for fuelwood or pots broken in a fight at a public well. They pit rich against rich as in 1991, when Bangarappa and Jayalalitha, the Chief Ministers of Karnataka and Tamil Nadu respectively, engaged in a slanging match and precipitated violence in a dispute over the use of Kaveri waters for the rich farmers and city dwellers of the two states.

LAND TO THE TILLER

In this chapter we propose to describe a range of such conflicts pertaining to access to and control over natural resources. In an agrarian country such as India the most basic resource is cultivable land, a resource that has become all the more critical as village-based crafts have been collapsing one by one (a process that began with the deliberate sabotage of the Indian weaving industry under British rule, in order to open up markets for the mills of Manchester). This process has continued after independence, with government policies relating to synthetic yarn production and wool weaving further undercutting the viability of a cottage industry that once employed a very significant proportion of India's population (Jain 1983).

A fundamental failure over large parts of India has been that of land reforms. While all land in pre-British India belonged in theory to the crown, peasant communities controlled it very effectively. With the land : person ratio still quite favourable, peasants had the option of moving somewhere else and starting cultivation if a local ruler became particularly oppressive and demanded too large a fraction of surplus agricultural production. Agricultural land was not a commodity commonly bought and sold, so that a peasant could not easily be alienated from his land. Local chieftains collected the surplus of agricultural production, usually in kind, and passed on a fraction to the king. A good bit of the land was also assigned to temples whose controlling priests appropriated the surplus. In this system too an omnivore class of warriors, priests and bureaucrats lived on production appropriated from ecosystem people. But the latter still enjoyed a substantial degree of autonomy and control over their local resource base.

The British land settlements helped tie peasants down to particular pieces of land, demanded high levels of taxes in cash, converted agricultural land into a commodity that could be bought and sold, and created a large class of tenant farmers cultivating on rent land that was owned by a handful of landowning families. This was done more directly in the eastern and northern territories, which were conquered first, with the so-called *zamindari* settlements assigning all ownership rights over large tracts of land to landlords. It was achieved indirectly in the western and southern territories conquered later, when peasants unable to pay the heavy land revenue assessments in cash got into debt and lost their lands to moneylenders, who were chiefly from trading and priestly communities. The colonial landholding systems created a large body of impoverished peasantry, perpetually in debt, in the Indian countryside. Its numbers swelled as large numbers of rural artisans such as weavers, nomadic traders and river ferrymen lost their base of subsistence owing to the import of British goods and improvements in transport and communication. Their numbers grew further after the 1920s as the population, which had declined with a series of disastrous famines and epidemics between 1860 and 1920, assumed an upward trend.

Throughout British rule this mass of ecosystem people remained dependent on a largely stagnant system of agricultural production, many tilling land as sharecroppers in perpetual debt. They backed the national struggle for independence expecting fully that the land would come to the tiller after the British left. Indeed, the Indian National Congress promised such a reform; but the promise has been left substantially unfulfilled, especially in the *zamindari* territories (with the notable exception of the communist-ruled state of West Bengal), where the British had installed landlords on a large scale. Land reform has gone much further in territories where land was initially assigned to the peasants, though many of them later lost it to moneylenders. In these *ryotwari* states of Maharashtra and Karnataka, for instance, land reform has progressed substantially.

The political response to this failure has been twofold: the Gandhian, as represented by Vinoba Bhave and Jai Prakash Narayan, and the socialist, as represented by a whole spectrum of groups from social democrats to revolutionary communists. The Gandhian approach has been largely ineffective. While large amounts of land were voluntarily donated, even whole villages in the early days of the Bhoodan and Gramdan movements, the Gandhians never got down to organize rural society on a new, egalitarian basis. The leftists on the other hand have proved far more effective, albeit on a still limited scale. It is the communist governments of Kerala and West Bengal that have pushed through relatively complete land reforms. In other states like Karnataka and Maharashtra too it has been the pressure from the left that has been instrumental in persuading Congress governments to implement land reform.

The failure to implement land reform over much of the country has fuelled violent leftist movements, beginning with the communist-led peasant revolts of Telangana in the 1940s. More recently, groups working for armed struggle have come to be known as Naxalites, after the uprising in 1967 at Naxalbari in West Bengal. This extreme-left movement has had considerable influence in the tribal areas of West Bengal, Bihar, Madhya Pradesh, Orissa, Andhra Pradesh, Maharashtra and parts of Tamil Nadu and Karnataka. The Naxalites have been instrumental in tribal and poor landless farmers staking claim to agricultural land and forests for cultivation (Naidu 1972; Calman 1985). Although they do control some areas, and even extort taxes (for instance from forest contractors in Bastar), Naxalites have had little direct success in promoting land reforms or local control over forest resources. But their influence has undoubtedly been of significance in the recent moves towards greater involvement of local people in forest management (the history of agrarian movements in twentieth-century India is usefully surveyed in the volumes edited by the sociologist A.R. Desai (1979, 1986)).

This failure to implement land reforms, especially in the fertile agricultural tracts of the Gangetic plains, has meant that enhancement of agricultural production cannot be a broad-based effort. Instead, it has been pursued as an attempt to inject large quantities of outside inputs: water, fertilizers, pesticides, high-yielding varieties over limited areas to produce large surpluses of agricultural production that can be siphoned off to support a growing urban population and agro-based industries such as sugar and cigarettes. This is the 'green revolution', which has primarily occurred in the northern states of Punjab and Haryana. These are areas of relatively low rainfall where pastoralism had been the mainstay of the traditional economy, supplemented to a limited extent by agriculture. Since pastoralism, especially when it involves nomadic movements, is difficult to tax, no major landlords emerged in this region, which initially yielded little revenue for the British government. But the soils of these states are fertile alluvium and there was considerable scope for irrigated agriculture if the tributaries of the Indus and the Ganga could be dammed. A major thrust of British policy therefore was to bring these lands under irrigated agriculture. This was continued more vigorously after independence, converting Haryana and Punjab into the agriculturally most productive areas of modern India, along with parts of the Godavari, Krishna and Kaveri deltas.

But this strategy of boosting agricultural production in limited areas while permitting agriculture to stagnate on fertile lands owned by conservative landlords elsewhere has meant that the masses of ecosystem people of the Indian countryside remain impoverished, while a new class of omnivores has been created from among the large and medium-sized landholders of the areas of 'Green Revolution' agriculture. This is how the landless of Bihar have become migrant agricultural labourers in Punjab, where they have been massacred time and again by terrorists in yet another manifestation of social conflict flowing from inequitable development. This lopsided agricultural development is also accompanied by a series of undesirable environmental consequences, including waterlogging of overirrigated lands, depletion of soil fertility, pesticide pollution, nitrate pollution of the underground aquifer and so on.

The enhancement of agricultural productivity over the past three decades has thus left much of India's rural population out of its ambit. Nor has there been a significant enough increase in employment in the industrial sector. Together this has spawned enormous numbers of people in the Indian countryside with a hunger for ownership of land for cultivation. In the absence of land reform, these people have no recourse but to encroach on common lands, be they village grazing lands controlled by revenue authorities or reserved forest lands. Such pressures have led to a variety of endemic conflicts, each on a small scale, but brewing across the length and breadth of the country. Thus landowners wish to maintain the grazing grounds intact, while landless agricultural labourers, mostly of the lower castes, attempt to occupy these for cultivation. Perched atop the Nilgiri hills at an altitude of 2,400 m is a village called Nanjanad. Most of the village land here is owned by a community of commercially oriented farmers called *Badagas* who have been cultivating potato and tea. Just outside the village is a hamlet of scheduled caste labourers who were working on *Badaga* fields. In 1990 several among them occupied the village grazing lands, in part encouraged because the then District Collector was actively assigning land to lower-caste families. The *Badagas* rose as one and ensured that they were evicted, and the district collector transferred. Similar skirmishes have been reported from many other parts of the country as well.

With enormous inputs concentrated in urban areas, the value of urban land is far higher than that of agricultural land, even if the latter is irrigated. Buildings on urban land require bricks, which can be formed from the soil of agricultural fields, baked with wood from village trees. This makes a quick profit, although it drastically reduces agricultural yields. So on the fringes of all urban centres in India, old banyan and peepul trees, once revered as sacred and therefore never to be cut, or mango trees previously maintained for their fruit, are firing kilns with bricks made of topsoils from the field (Gadgil 1989). There are sporadic protests against these trends, as near the town of Honnavar in coastal Karnataka, where a voluntary agency called Snehakunj is trying hard to stop brick and tile factories from buying the topsoil of paddy fields.

With the price of urban land skyrocketing, there is a scramble to acquire pieces of it. New packages of urban land are continually created by the various urban development authorities, with the land being made available not only below the value commensurate with state inputs into its development, but even below the prevalent market prices. This creates a scramble for such land and an opportunity for politicians and bureaucrats to extend patronage while making a killing. There are scandals around land development in most major cities with accusations of corruption against politicians of virtually all parties. One such celebrated case pertains to housing development proposed in land surrounding the Thippegondanahalli reservoir, which provides 30 per cent of the total supply of drinking water to the city of Bangalore. Land around this reservoir was purchased by forty-two ex-army officers in 1972, when it did not come under the purview of the Land Reforms Act. A special commission granted conversion of land for non-agricultural purposes in 1979 after permission was obtained from the Pollution Control Board and the town planning authority to construct 771 houses. In 1982, however, the Bangalore Water Supply and Sewerage Board appealed against the project on the grounds of the threat of pollution to the reservoir. In response, a firm of Delhi-based promoters put up a proposal for construction of 270 country villas in 414 acres (168 ha) of land around the reservoir, arguing that villas, instead of group housing, would protect the green belt. A single judge quashed the order permitting the formation of the layout in 1987 and this judgment was upheld in 1991 by a division bench of the high court. During Mr Bangarappa's chief ministership of the state of Karnataka a government order was passed approving the formation of a township around the reservoir in contravention of these court judgments. Opposing this order, A.R. Lakshmisagar, a former Law Minister of the state, and several others filed a public-interest litigation. Questioning the validity of the government order, they submitted that this constituted contempt of court and argued that with the formation of the proposed township not only would the quantity of water flowing into the reservoir deplete, but the water would be contaminated as construction workers would flock into the area, forming slums around the reservoir. They also pointed out that the order had been passed in great haste, it having taken only ten days for the files to move from one department to the other. Mr B. Basavalingappa, Minister for Environment and Ecology in Mr Bangarappa's government, also strongly condemned the government order. He accused the promoters of processing their application through the Secretary of Housing and Urban Development, who had nothing to do with the clearance of the project. The high court now has set aside the government order approving the formation of the township, branding the state's action as 'arbitrary and high-handed'. The division bench of the high court remarked that by overruling the previous high court judgment, the government had not only thrown the laws to the wind, but also violated the writs issued by the court. In addition, the high court pointed out that the government order had also contravened the Land Revenue Act.

In all this, ecological refugees flocking to the cities have little chance of proper housing given the enormous prices of urban land. So they create shantytowns on land peripheral to the city, without any organized transport, water or power supply. As the numbers of people in such slums mount – and they have come to constitute as much as half the population of the metropolis of Bombay – they form significant vote banks. Political bosses, often doubling as slum lords, then help organize some water, power, transport facilities to such settlements. But as cities expand these once peripheral lands become highly attractive for developing higher-income housing or office buildings and acute conflicts may develop over clearance of slums, so that omnivores can gain fuller control over such land. A major violent incident during the Emergency years of 1975–7 thus involved demolition of slums in Delhi, and real estate developers are suspected to be involved in the Bombay communal riots of 1992–3 which led to large-scale destruction of slum areas.

DAMS AND THE DAMMED

Along with land, water is the resource in widest demand. Some of the most virulent conflicts have arisen when omnivores attempt to capture water resources by denying ecosystem people access to cultivated land. A major context in which ecosystem people are thrown off their lands is the construction of major dams – the most ambitious of which are the chain of dams being built on the River Narmada. The movements of dam-displaced people have gathered force in the past two decades. We shall come to these contemporary protests presently, but we must first note one important, though as yet little known, precursor. Known as the Mulshi *satyagraha*, this was the opposition to a dam being built on the Western Ghats south of Bombay by the flourishing industrial house of the Tatas. This episode is virtually unknown to Indian environmentalists – and in view of the remarkable parallels between the Mulshi *satyagraha* and ongoing protests against large dams, its history is worth recording at some length.

The Tatas had in fact planned an ambitious series of dams on the Sahyadri hills, chiefly to supply power to the rising industrial city of Bombay. When the first dam was built near the hill station of Lonavala, the farmers and herders whose lands were submerged were paid no compensation whatsoever. But when the Tatas came to Mulshi for the next phase of the project, they ran into trouble. At first, the company moved on to the farmers' lands and dug their test trenches without any legal formality. But Mulshi was very close to Pune (Poona), then an epicentre of the Indian freedom movement. So when a peasant objected to a trench being dug in his field and a British engineer threatened him with a pistol, there were strong protests in Pune.

The ensuing opposition to the dam was led by a young congressman, Senapati Bapat. Bapat and his followers succeeded in halting construction of the dam for a year. The Bombay government then promulgated an ordinance whereby the Tatas could acquire land on payment of compensation. Now the resistance to the dam split into two factions. Whereas the *Brahman* landlords of Pune, who owned much of the land in the Mulshi valley, were eager to accept compensation, the tenants and their leader, Senapati Bapat, were totally opposed to the dam project. With the landlords, the power company and the state all ranged against them, there was little the peasants could do, and the movement collapsed in its third year. Tragically, the compensation was pocketed by the landlords, and the actual tillers of the soil were left high and dry. However, the movement at least succeeded in forcing the Tatas to provide reasonable negotiated compensation for the submerged lands. In consequence, they did not proceed with the other hydroelectric projects they had intended for the Sahyadris (these were later taken up as state-sponsored projects on independence, with the displaced people still ending up greatly impoverished).

When the Mulshi *satyagraha* broke out, the British District Collector had toured the area, extolling the virtues of the dam. He remarked that the electricity produced by it would light up the latrines of the Bombay *chawls*, the dwelling homes of the city's industrial workers. This drew the sharp retort that the government and the Tatas sought to extinguish wick lamps in thousands of rural homes to light up the latrines of Bombay (Bhuskute 1968).

This exchange, apocryphal as it might be, could well have taken place in 1990 in either Ferkuva or New Delhi between proponents and opponents of the Sardar Sarovar dam. In fact, when the Narmada controversy was at its height, *The Times of India*, whether by accident or design, reproduced in its archive section a report, dated 2 May 1921, on the Mulshi *satyagraha*. Here the paper's correspondent had succinctly represented the main objections to the Tata project, as well as its most powerful justifications. The origins of the Mulshi *satyagraha*, he remarked, lie in:

1 A strong sense of wrong and deep feeling of resentment among the peasantry whose lands are affected by the project, against the government for sanctioning the scheme more than two years ago, without taking them in its confidence, i.e., without consent, knowledge or consultation of the peasant-owners of the land.

2 Suspicion and distrust in both the government and the company, due chiefly to the procedure of acquisition, as to the bonafides of their intentions to award full compensation, or equivalent . . . land somewhere else, and other facilities already enjoyed by them or necessary for fresh colonization.
3 Reluctance to part with the land on account of its extreme productivity, the natural facilities of irrigation and nominal amount of land revenue.
4 Reluctance to part with lands, ancestral homes, and traditional places of worship and see them submerged under water.
5 Natural reluctance in this class of peasantry to emigrate from one place to another.

On the other side the main claims of the project promoters were:

1 One and a half lakh [150,000] [of] electrical horse-power would be created by the Mulshi Peta dam.
2 It would save 525,000 tons [of] coal every year. This quantity of coal at the present rate costs Rs 18,300,000.
3 The saving of coal means a corresponding saving of Rs 10,750,000 worth of fuel to the mill industry of Bombay.
4 The quantity of coal saved on account of the scheme would require 26,250 wagons for transport. These would be saved and utilized for other public purposes.
5 Water once used can be directed for agricultural purposes after electrical power is created.
6 Electricity thus created would give work to 300,000 labourers. If it is utilized for cotton mills, every day 51 lakh [5,100,000] yards would be manufactured.
7 The projected electrification of the Bombay suburban railway lines would give to Bombay city much faster and more frequent trains, thus enabling the development of housing schemes in purer air and healthier circumstances.

Here lies an uncanny anticipation of the ideological roots of the conflicts over large dams that were to erupt half a century later. On the one side, the interests of subsistence-oriented peasants; on the other, the interests of urban centres and industry. When the major push towards river valley projects took place after Indian independence, it was easy to represent the former as static and backward, the latter as dynamic, forward-looking and coterminous with the national goals of progress and development. The villages to be submerged by the new projects were then expected to make way for the larger national interest, the more so as the new schemes (unlike Mulshi) were owned and executed not by private capitalists but by the state itself, the legatee of a broad-based, popular national movement.

Of course, displaced people were not entirely yielding. When the foundation stone of the Hirakud dam in Orissa was laid in March 1946, there were strong protests by the peasants of Sambalpur district who were to be ousted by the project. The dam was to inundate fertile rice tracts as well as substantial reserves of coal and other minerals. When the dam site was visited by the Public Works Department Minister, he was confronted with a demonstration of 4,000 people. Beginning in September 1946, several processions and strikes were organized in Sambalpur town. Section 144 of the criminal procedure code was served, forbidding large gatherings. Defying prohibitory orders, on 12 November a strong procession of 30,000, including many women, marched through the town shouting slogans against the dam. However, the agitation fizzled out shortly afterwards, in part due to the co-option of its leaders by the Congress and the administration (Pattanaik *et al.* 1987; Misra 1946).

A comprehensive if somewhat euphoric survey of the first wave of large dams built in independent India, by the political scientist Henry Hart, also noted the resentment of villagers confronted with the prospect of displacement. Thus in 1953, the residents of the town of Narayan Deva Keri, in present-day Karnataka, hoped desperately that the new reservoir on the Tungabhadra river would not fill up to capacity, thereby sparing their town. Disregarding the warnings of engineers, the townspeople stayed on until the last moment, having to be evacuated in a hurry when surrounded on three sides by water. Despite these signs, there was general agreement, at least among the votaries of dam building, that 'the suffering of the displaced people was for the good of the greatest number', as well as little doubt of the 'willingness of the Indian villager to make way for a nation building project, provided he is convinced that the sacrifice he is called upon to make is unavoidable' (Hart 1956: ch. 12).

It is true that the massive – one might, following Hart, call them heroic – river valley projects of the 1950s met with little opposition. These included the Bhakra-Nangal dam in Punjab, the Tungabhadra project on the Andhra Pradesh–Karnataka border and the Rihand dam in Uttar Pradesh, each displacing tens of thousands of people. And yet, over time the Indian villager was to develop a marked unwillingness to make way for 'nation-building' projects. A major reason for this growing hostility was the actual experience of communities displaced by earlier projects. The resettlement of dam evacuees has uniformly been inadequate: the rates of cash compensation have been low; the promise of land for land has very rarely been fulfilled (and where it has, the new lands are invariably of much poorer quality); not to speak of the difficulties of making a new home in unfamiliar, and often hostile, surroundings (see, for a review, Centre for Science and Environment 1985; Fernandes and Ganguly-Thukral 1988; Ganguly-Thukral 1992). Indeed, the experience of dam oustees in India validates the grim judgement of the anthropologist Thayer Scudder that 'next to killing a man, the worst you can do is to displace him' (quoted in Singh 1994: 259).

A significant acknowledgment of these failures has been the substitution, in recent years, of the term 'displacement' for the euphemistic 'resettlement' in public discussions of this process. Meanwhile, organized opposition to new projects gathered force in the early 1970s, with movements emerging independently in different parts. The most long-standing opposition has been to the Tehri dam, being built on the River Bhageerathi in the Garhwal Himalaya. For a decade and a half, the dam's construction has been opposed by the Tehri Baandh Virodhi Sangarsh Samiti (Committee for the Struggle against the Tehri Dam), a forum founded by the veteran freedom fighter Virendra Datt Saklani. The respected Chipko leader Sunderlal Bahuguna – whose own *ashram* is not far from Tehri town – has been very active in the movement, undertaking several fasts to pressurize the government to stop construction. The objections to the dam relate to the seismic sensitivity of the fragile mountain chain (hence the possibility of a dam burst), the submergence of large areas of forest, agricultural land and the historic town of Tehri, and the threat to the life of the reservoir owing to deforestation in the river catchment (D'Monte 1981; Valdiya 1992). These criticisms have gathered force after the massive earthquake in the upper Bhageerathi valley in October 1991, but the government appears resolved to go through with the dam.

Simultaneously, the other well-known Chipko leader, Chandi Prasad Bhatt, has been leading the resistance to a dam at Vishnuprayag, on the Alakananda river in eastern Garhwal. This construction is taking place very close to the famous Valley of Flowers, and fears that the ecology of the valley would be permanently disturbed are compounded by the geological features of the Vishnuprayag area, one peculiarly prone to landslides (Bhatt 1992). At the time of writing, the Vishnuprayag project, in part owing to such opposition, has been indefinitely put on the shelf.

The participation of Chipko activists in these protests is hardly accidental. Having largely lost their forests to commercial exploitation, Himalayan peasants now face further suffering owing to external pressures on the other resource their hills are abundant in, water. As in the case of forests, the benefits of intensive water exploitation have gone almost exclusively to the inhabitants of the plains. In a comparable fashion, the water-rich and heavily forested tribal areas of central India have also witnessed a surge of opposition to new hydroelectric projects. Three of the more notable movements have arisen in opposition to the Koel Karo and Subarnarekha dams in Bihar, and the Bhopalpatnam-Inchampalli project on the Maharashtra–Madhya Pradesh border. In all these cases, threatened tribal groups have responded spiritedly to defend their homes, by organizing demonstrations and work stoppages – all this in the face of police harassment, beating and other forms of repression by the state (Centre for Science and Environment 1985; Areeparampil 1987).

However, the groups affected by large dams have not always been tribal in origin. One successful movement was in fact led by prosperous orchard owners in the Uttara Kannada district of Karnataka. Here the Bedthi project was abandoned after it was opposed by influential spice garden farmers, largely *Brahman*, whose lands were to be submerged by the project. The Uttara Kannada farmers succeeded in organizing an alternative study and the country's first public appraisal of a development project. After hectic lobbying with political leaders, they managed to force the state government to abandon the dam (Sharma and Sharma 1981).

Another, and more striking, success was the abandonment of the Silent Valley hydroelectric project in the state of Kerala. No human community was to be displaced by this 120-kW dam, which nevertheless did intend to submerge one of the best surviving patches of rain forest in peninsular India. The opposition to the project was led by the Kerala Sastra Sahitya Parishad (KSSP), an organization dedicated to popular science education that has a wide reach and influence in Kerala. This left-leaning movement of school and college teachers here built up an unlikely collaboration with wildlife conservationists (Zachariah and Sooryamurthy 1994). Each group has its own reasons for opposing the project. While the KSSP emphasized the techno-economic appraisal of energy-generating alternatives, their allies invoked the need for plant and animal conservation. Eventually, Mrs Indira Gandhi's desire to enhance her image among the international conservation community appears to have been critical in the central government's decision to shelve the project (see D'Monte 1985).

There is, then, a considerable history to the movement against dam construction on the Narmada river. The Narmada river valley project – which the writer Claude Alvares has termed the 'world's greatest planned environmental disaster' – is a truly Gargantuan scheme, envisaging the construction of thirty major dams on the Narmada and its tributaries, not to speak of an additional 135 medium-sized and 3,000 minor dams (see Kalpavriksh 1988).

With two of the major dams already built, the focus of popular opposition has been the Sardar Sarovar reservoir, the largest of the project's individual schemes. Sardar Sarovar is unique in the history of dam building in India, in that the command area of major beneficiaries lies in one state, Gujarat, while the major displacement will occur in another state, Madhya Pradesh (of the 243 villages to be submerged by the dam, 193 lie in Madhya Pradesh). According to official estimates based on the already outdated 1981 census, over 100,000 people, of whom approximately 60 per cent are tribal, will be rendered homeless (Vinod Raina, personal communication).

As early as 1977, villagers in the Nimad region of Madhya Pradesh began protesting against the prospect of eviction due to Sardar Sarovar. Notably, social activists like Medha Patkar (now one of the Narmada Bachao Andolan's moving spirits) first began work on the proper rehabilitation of potential oustees. The government of Maharashtra had by then declared a very progressive 'land for land' policy for oustees of dams. This policy called for allotment of an equivalent amount of good land for the land being submerged.

But given India's population pressure, nearly all such land is already under cultivation or habitation. The most appropriate (and just) solution would, then, be to acquire surplus land from the larger landholders in the command areas of the dam. These landholders are the prime beneficiaries of investment by the state in enhancing the productive potential of their lands, investments to the tune of Rs 52,000 per ha. But these are people with political clout, and the state machinery is not motivated enough to acquire their surplus lands for distribution among displacees. It was therefore only on realizing that there was no land available in Madhya Pradesh, Maharashtra or Gujarat for the proclaimed 'land for land' policy that Medha Patkar and her colleagues turned to opposing the construction of the dam itself.

Although more than a decade old, the movement has really gathered force only since 1989. Over the past five years it has used a varied repertoire of protest to put forward its demands: *rasta rokos* (the blockade of roads and their traffic); public meetings (including some where oustees have pledged not to leave their homes even if the dam waters rise and drown them); fasts; and demonstrations, especially at state capitals. In one dramatic incident, villagers from the neighbourhood of Badwani town in Madhya Pradesh uprooted stone markers from the dam's submergence area, transported them several hundred kilometres to the state capital of Bhopal and flung them outside the Madhya Pradesh legislature (see *Narmada* 1989–90).

While localized protests have been occurring all along the Narmada valley, wider public attention has been drawn through spectacular events. Two of these have already been mentioned: the congregation outside the Prime Minister's house in New Delhi, and the protest march from Rajghat to Ferkuva. Yet the most successful of these public events was a great rally in the town of Harsud, held on 29 September 1989. Sixty thousand volunteers, mostly of tribal and peasant background, had gathered in Harsud, a town itself destined to come under 15 m of water. Representatives of citizens' groups from all over India also came to demonstrate their solidarity with the Narmada movement. A large public meeting, addressed by Amte, Patkar, Bahuguna and others, culminated in a collective oath to resist the pattern of 'destructive development' exemplified by the Sardar Sarovar dam (Alvares 1989).

Even though it lies in a path of continuity with them, there are several features which help distinguish the Narmada movement from earlier protests against large dams. Two of the most notable are its spread – it has activist groups working in three states and many supporting organizations elsewhere – and its tenacity in the face of government repression. Although the movement itself has been almost completely non-violent, its leaders and participants have been repeatedly harassed, and occasionally beaten and jailed. Again, unlike many other movements, the Narmada Bachao Andolan has been widely, and often sympathetically, covered in the print media, while it also has well-established links with environmental groups overseas. (Japanese environmentalists have persuaded their government not to advance money for the Narmada valley project, while US groups sympathetic to the movement tried hard to convince the World Bank to do likewise.) A final testimony of the movement's vigour is the active counter-movement of omnivores it has generated in support of the dam. Political leaders and social workers in Gujarat have strongly rallied behind the state's rich farmers – who, along with the building contractors, stand to gain most from the project – organizing demonstrations and press campaigns and mounting an ideological offensive wherein the Narmada movement's leaders are portrayed as 'anti-development' and 'anti-national'. The Narmada activists have even been accused of wanting to deny tribals the fruits of economic growth by keeping them in a perpetual state of nakedness, hunger and illiteracy (see Anklesaria Aiyar 1988; *Economic and Political Weekly* 1991).

One most disturbing aspect of omnivore opposition to the Narmada Bachao Andolan (NBA) has been growing state repression. Thus on 29 January 1993, two hundred policemen entered the tribal hamlet of Anjanvara, in the Jhabua district of Madhya Pradesh. Anjanvara was one of the villages which had been resisting government surveys conducted preparatory to submergence. The tribals had been organized by the Khedut Mazdoor Chetana Sangath, a local activist group active in the wider Narmada Bachao Andolan. The policemen who entered Anjanvara beat up some villagers, ransacked their houses and fired eight rounds. Three days later the police mounted another attack, following which they arrested twenty-one villagers.

Nine months later, in November 1993, tribal protesters in the Dhule district of Maharashtra found themselves at the receiving end of police brutality. In the village of Chinchkhedi, police fired forty-six rounds on a demonstration of tribals, killing a 15-year-old boy, Rehmal Punya Vasava. As in Jhabua, the tribals had themselves been militant but unarmed. In the most recent incident of this kind, the Gujarat police refused to intervene when political hoodlums broke into the NBA office in Baroda, beating up social workers and tearing up files.

For all this intense repression, at the time of writing (October 1994) the Narmada controversy is far from resolved. Thus in March 1993, the Indian government withdrew its request for a loan for the project from the World Bank – a pre-emptive move before the Bank itself was likely to have decided, on the basis of an adverse report by a review committee it had appointed, not to support the project (*The Times of India*, 31 March 1993). Some months later, in Manibeli village, Maharashtra, which was destined to be submerged by rising waters during the monsoon, a group of tribals and NBA activists resolved to drown rather than move out. They had to be forcibly removed from Manibeli by a police force. Then in June 1993, Medha Patkar went on a fifteen-day fast in Bombay, which attracted much attention and support. Following her fast, Patkar and her colleagues met with the union government's Water Resources Minister, who promised them a full review and reassessment of the project. Most recently, in March 1994 the new Congress Chief Minister of Madhya Pradesh, Digvijay Singh, proposed a reduction in the height of the dam, so as to lessen the burden of displacement. But his proposal has been dismissed by the government of Gujarat, and construction work continues on the dam.

In areas beyond the Narmada basin, the ecosystem people are losing their control over land in the context of other development and defence projects as well. Among the more notorious of these is the Baliapal missile range project in the Balasore district of Orissa. In the delta of the Subarnarekha River, the Ministry of Defence proposed to build a huge test range for missiles, a project that envisaged the acquisition of 190 sq km of land and would have displaced an estimated 70,000 people. But this is fertile land, with a highly developed agrarian economy based on betel leaves, coconut and fishing. In 1987 and 1988, a popular movement against the test range gathered force in the Baliapal-Bhograi region. The contempt of the peasants for the government was most dramatically manifested when they barricaded all entrances to the region, refusing to pay taxes or allow the entry of government officials. This movement forced a stalemate: the government has not yet been able to proceed with the project, and there has been talk of shifting it elsewhere. The general secretary of the People's Committee against the Test Range succinctly expressed the feelings of the people of Baliapal, and indeed of all those threatened with displacement by large projects, when he remarked: 'No land on earth can compensate for the land we have inherited from our forefathers' (*Indian Express*, 24 April 1988; *Deccan Herald*, 15 May 1988).

A Baliapal-type situation is currently being replicated in Orissa's neighbouring state of Bihar. Here, in the hilly, predominantly tribal and desperately poor district of Palamau, the Indian Army proposes to build a massive test firing range. Stretching over 40,000 ha in Palamau and the adjoining Gumla district, the construction of the range will affect a total of 190 villages. The firing range might also have a powerful negative impact on the biodiversity of the Betla National Park, a Project Tiger reserve. But here too the ecosystem people are resolved not to give up their lands and forests without a fight. Youth activists have formed a *jan sangharsh* (popular struggle) committee, and impressive protest demonstrations have been arranged in the towns of Ranchi, Gumla and Daltonganj. As one tribal told a visiting journalist, 'Hum jan denge lekin zamin nahin denge' ('We shall give up our lives, but not our land') (Sainath 1993).

WATER AS SOURCE AND SINK

Water, along with land, is the most vital input for agricultural production. In a country in which the majority of people depend on agricultural production, access to water is of critical significance. But water is an equally important resource for many industries, at least as a sink for their waste products.

Drinking water is important for rural as well as urban settlements. Control over water is therefore a major source of social conflicts in India and inequitable control leading to mismanagement of water resources underlies many aspects of India's environmental crisis.

Water comes down as rain. It flows on the surface as well as under the ground. While access to surface water can be readily controlled and manipulated, ground water may be tapped at will by the landowner, so long as he or she has the resources to reach down and lift it. When the rate at which the ground water is withdrawn exceeds the rate at which it is being recharged, the level of the ground water table drops, so that more and more effort is required to get at it. Indians have been tapping ground water through open wells since the days of the Mohenjadaro civilization; but the rate at which this water could be extracted was previously limited by the capabilities of human and bullock power. This limitation has now been overcome with the development of diesel or electric power operated pumps, so that water below the surface can be pulled up at rates far exceeding those of recharge, leading to a rapid lowering and even local exhaustion of ground water. This opens up possibilities for conflict, with those capable of reaching down further and deploying more powerful pumps depriving others of access to ground water. With the state stepping in to subsidize digging of bore wells and purchase of pumps, and providing electrical power at nominal rates, over much of the country the better-off farmers have benefited at the cost of others.

This process is well illustrated in an exhaustive recent study of the ground water situation in rural Gujarat. Over a twenty-five-year period, there has been a nearly sixtyfold increase in the number of electrified pump sets and tube wells in the state. The interests of rich farmers have fuelled an indiscriminate expansion of water extraction devices and of water-intensive crops. These rural omnivores have been aided, in the tiresomely familiar pattern, by liberal financial assistance from the state and by greatly subsidized electricity. As a consequence, the water table has fallen alarmingly – in many parts of Gujarat, by several metres or more. This means that smaller peasants often no longer have access to water for their crops, for the ground water level is now well beyond the traditional dug wells, which are all that they can afford. Ironically, while the state as a whole has been plagued by water scarcity and drought, the rural omnivores are actually able to profit from drought conditions, owing to their near monopoly over water. Thus the ground water economy of Gujarat has been summed up in those two words, *overexploitation* and *inequity*, that so accurately capture the overall process of natural resource development in independent India (Bhatia 1992).

Another illustration from our own experience comes from the coastal Kumta *taluk* of the state of Karnataka. Here there are many villages where the richer, upper-caste *Haviks* own betelnut-cum-pepper orchards, while the poorer, lower-status *Halakki Vakkals* till small paddy fields. Prior to the 1960s, when water could be lifted only with the help of muscle power, the orchards were lightly irrigated and produced modest yields. The water table remained quite high, so that the *Halakkis* could grow two crops of rice a year. But then the infrastructure for lifting up water using electrical power became readily available and the better-educated, richer *Haviks* growing cash crops monopolized state subsidies. They began to irrigate their orchards more intensively, got higher yields, and as prices of betelnut and pepper rose came to accumulate more and more wealth and power. The rice-cultivating *Halakkis* found no entry into this system. Instead they were at the receiving end: unable to grow a second crop of rice after the monsoon as the water table plunged down, and further impoverished as the rice prices did not keep pace with the rate of inflation. The result has been growing tension between *Halakki* and *Havik*, a conflict tinged with a communal element.

Large-scale irrigation works are not new to India. Mauryan emperors were already constructing dams, and an extensive Yamuna irrigation canal was functioning in Mughal times. But larger-scale irrigation as well as hydro-electric projects really began in the British period, and took off as the most prominent ingredient of the development effort following independence. We have already recounted the struggles over land required for the construction of these projects. There are equally portentous struggles over water, especially acute because those who do get the water pay so little for it. The beneficiaries are in effect being enormously subsidized by the state exchequer and fight bitterly to corner these benefits.

Among the more celebrated of such conflicts is that over the sharing of the Kaveri waters between the omnivores of the states of Karnataka and Tamil Nadu – be they sugarcane farmers of Mysore, urban dwellers of Bangalore, industrialists of Mettur or rice growers of Tanjore. The Kaveri originates in Karnataka and Kerala and joins the sea in Tamil Nadu, being dammed several times on the way. The contention of Karnataka is that the formula for sharing of water arrived at under British rule was unfair, since the Karnataka stretch of the Kaveri was then under the native state of Mysore while the Tamil Nadu portion came under direct British rule. Tamil Nadu naturally insists on a continuation of this agreement. In Karnataka this led in December 1991 to *bandhs* (shutdowns) and counter-*bandhs* and mob violence while the state-controlled police watched idly. The chief victims of this violence were the poor migrant Tamil labourers and small cultivators, ecological refugees who had come to Karnataka to eke out a livelihood.

The state invests enormous amounts, often raised through World Bank loans, to bring water to the cities and distribute it practically free of cost to city dwellers. There are no firm computations of the extent of subsidies, but the citizens of Bangalore pay no more than 5 per cent of what it has cost the state to deliver water to them. The investment to Karnataka in bringing water to those 10 per cent of the citizens of the state concentrated in Bangalore is, in order of magnitude, greater than the investment for water supply to all the other villages, towns and cities of the state put together. In spite of this,

Bangalore, which is located on a high point on the Deccan plateau at an altitude of nearly 1000 m, and away from all the river courses, is chronically short of water. So while the five-star Ashoka Hotel at the highest point in Bangalore has running hot and cold water on tap for luxurious tub baths, and rich households can fill huge underground sumps, the poor must fight to fill a few pots at public taps which are dry except for a couple of hours every other day. The quarrels at these taps, along with the struggle to put up a hut made out of old tar drums in crowded slums, symbolize the internecine strife that India's ecological refugees are tragically involved in today.

With water a free good, it makes perfect economic sense for a private entrepreneur to pollute his surroundings instead of investing in technology to properly treat and safely dispose of effluents. The state, as the repository in theory of the welfare of the public, then emerges as the agency most likely to pass legislation to check pollution and take punitive action against offenders. Indeed, in the industrialized world a major focus of the environmental movement has been on pressurizing the state to pass legislation and create enforcement agencies to check air and water pollution (see Hays 1987).

In India too, pollution control legislation is on the statute books, but because the gulf between omnivores who impose the cost and ecosystem people and ecological refugees who bear it is so large, industrial pollution has largely gone unchecked. In its executive functions, the Indian state apparatus alternates between being soft and predatory; in the first incarnation, laws are

Plate 14 There is very inadequate control of industrial pollution in the country

not enforced, while the second allows offenders to buy official compliance. Early in 1993, for instance, the Ministry of Environment and Forests of the Government of India yet again relaxed its schedule for gross polluting industries to clean up or face closure. It extended its earlier deadline of 31 December 1992 by one year for new ventures and by two years for ventures set up before 1981. It also reversed its decision to periodically monitor pollution, so that industries are now required to obtain a permit just once, except when planning expansion (*Down to Earth*, 15 March 1993). Yet, in a democratic political system, citizens' actions can act as a partial corrective even when the state abdicates its role.

Among the most notorious of industrial polluters are paper and rayon factories. Three units of Gwalior Rayons – owned by India's largest industrial house, the Birlas – have been indicted by environmentalists for affecting the economic welfare of downstream villagers through pollution. The Gwalior Rayon factory on the Chaliyar river in Kerala was closed for seven years after a spirited movement, led by the Kerala Sastra Sahitya Parishat. In the adjoining state of Karnataka, Harihar Polyfibres – owned by the same parent company – has faced concerted opposition and a long-drawn-out court case for discharging untreated effluents into the Tung-abhadra river. Villagers have complained of being hit by new diseases, declining fish yields and reduced availability of irrigation water (Hiremath 1987). A similar charge has been laid at the door of the massive Gwalior Rayons factory at Nagda in Madhya Pradesh, a state where the Birla-owned Orient Paper Mills, in the district of Shahdol, has also been criticized by social activists for its pollution of the Sone river (Roy *et al.* 1982).

Two other illustrations of the conflict between private profit and the public good come from Maharashtra, the western state with a highly developed industrial sector and a long tradition of social activism. In October 1987, farmers and fisherfolk from the Revadanda creek area of the Raigad district protested against the discharge of effluents from forty units operating in an industrial area owned by the Maharashtra State Industrial Development Corporation (MSIDC). Accusing the MSIDC of not treating effluents before discharging them into a nearby river, peasants inserted a wooden log into the discharge pipeline (*The Times of India*, 8 October 1987). Some months later, villagers in the Ahmednagar district of the state united to oppose the destruction of land and water by the discharge of south Asia's largest distillery. Despairing of remedial action, the villagers filed a suit in the Bombay high court, seeking Rs 10 million in damages from the offending unit, the Western Maharashtra Industrial Corporation, and the State Pollution Control Board (*Indian Post*, 8 August 1988).

Another example of citizen protest against pollution comes from the district of North Arcot in the southern state of Tamil Nadu. Here, effluents from a cluster of tanneries abruptly raised the chloride content of drinking water and contributed to declining crop yields by causing soil salinity. On

World Environment Day 1984, the town of Ambur, site of several tanneries, observed a total strike or *hartal*. Many women and children from the affected villages went in a procession through the town. Here women broke pitchers containing contaminated water, demanding that the authorities protect the health of their children. A huge effigy of an 'effluent monster' was also burnt on that day (*The Hindu*, 7 June 1984).

We have left until last what is perhaps the most tragic episode of the poisoning of the environment in human history. This took place at almost the exact geographical centre of India, the city of Bhopal, in December 1984. On the outskirts of the city lies the pesticide plant of the Union Carbide corporation, a US-based multinational. Early on the morning of 3 December there occurred an accident, allegedly caused by the introduction of water into the methyl isocyanate storage tank. This resulted in an uncontrollable reaction with the liberation of heat and the escape of methyl isocyanate in the form of a gas that killed, blinded or otherwise harmed humans. All around the Union Carbide plant was a shantytown full of ecological refugees. Within minutes of the gas beginning to escape, as many as 3,000 people were dead on the spot; altogether 50,000 were afflicted by chronic lung damage. But these numbers may err grossly on the low side, for there is a strong suspicion that the administration as well as the medical authorities greatly underestimated the magnitude of the disaster (Dhara 1992).

THERE'S TROUBLE IN FISHING

We now come to another category of ecosystem people whose dependence on living resources has also been undermined in recent decades. Distinct endogamous groups of specialist fisherfolk, both along the sea coast and on rivers, have long been a feature of the Indian landscape. These communities, which depend more or less exclusively on the catch and sale of fish, have recently been threatened by encroachments on their territory.

The problems of ocean-going fisherfolk have been well documented, particularly in the studies of the economist John Kurien. The clash between artisanal fisherfolk and modern trawlers, at its most intense in the southern state of Kerala, provides a chilling illustration of what can happen when one group's exclusive control over living resources is abruptly challenged by more powerful economic and political forces. For centuries, the coastal fish economy was controlled by artisanal fisherfolk operating small, unmechanized craft, who supplied fish to inland markets. In the 1960s, big business began to enter the fisheries sector. The advent of large trawlers, catching fish primarily for export, led to major changes in the ecology and economy of fisheries in Kerala. A rapid increase in fish landings in the early years of trawling was followed by stagnation and relative decline. While some artisanal fisherfolk were able to make the transition to a more capital- and resource-intensive system, the majority faced the full weight of competition

Plate 15 Rapid mechanization of fisheries has led to overexploitation of fish stocks and serious conflicts with artisanal fisherfolk

Plate 16 Mechanized fisheries have become an important component of the economy of the Andaman and Nicobar Islands

from trawlers. This conflict gave rise to a widespread movement – comprising strikes, processions and violent clashes with trawler owners – in which small fisherfolk pressed for restrictions on the operations of trawlers. The movement also called for a ban on trawling during the monsoon, the breeding season for several important fish species. A partial ban finally imposed in 1988 and 1989 did in fact result in an increased harvest following the monsoon months (Kurien and Achari 1990; Kurien 1993).

So far as inland fisheries are concerned, there have been, as illustrated above, intermittent reports of localized opposition by fisherfolk affected by industrial pollution. In a class of its own is a unique movement to 'free the Ganga' that arose among fisherfolk in the Bhagalpur district of Bihar. Here, in a bizarre relic of feudalism, two families claim hereditary rights of control over a fifty-mile stretch of the Ganga. Claiming that these *panidari* rights originated in Mughal times, the water lords levy taxes on the 40,000-odd fisherfolk living along the river. A protracted court case has so far been unsuccessful in abolishing these rights, which by an anomaly escaped the provisions of the acts abolishing landlordism (*zamindari*) enacted after 1947. Since the early 1980s, the fisherfolk have been organized by young socialists in the Ganga Mukti Andolan ('free the Ganga' movement). With fish catch also declining owing to industrial pollution, the movement has been waged simultaneously on two fronts – against effluents and an anachronistic system of monopoly rights over water (Narain 1983).

As India's natural fisheries are being depleted by siltation and damming of rivers, and by pollution and overfishing, the culturing of carp in freshwater ponds and shrimps in brackish water fields is becoming a lucrative business. Establishing control over these water bodies, especially where they earlier served as common property resources, can lead to conflicts on many different scales. In the Kumta *taluk* of coastal Karnataka is the estuary of the River Aghanashini. The extensive tidal mud flats of this river support the cultivation of a salt-resistant rice variety by the *Halakki Vakkal* community in the monsoon and serve as rich fishing grounds for the fisherfolk community of *Ambigas* in the dry season. In the late 1960s, however, these mud flats were embanked at government expense. Around the same time were set up seafood canning factories, so that shrimp came to fetch much higher prices. The bunds have permanently demarcated the paddy fields, with the result that the *Halakki Vakkal* farmers now claim ownership rights over shrimp fishing as well. This was contested by the *Ambigas*, resulting in litigation. The courts have ruled that *Halakkis* do control part of the estuarine land, but that deeper channels must be left open to fishing by the *Ambigas*. But incomes from auctioning of rights to shrimp fishing are high and the farmers are refusing to abide by the court ruling, continuing to keep *Ambiga* fisherfolk – who are far less educated and much poorer than the *Halakki* farmers – out of their traditional fishing grounds.

A very similar conflict, albeit on a somewhat bigger scale, has now erupted over the proposal to take up shrimp culture in Chilika lake, India's largest body of brackish water. Spread over 11,000 sq km in the state of Orissa, this huge lagoon is connected by a narrow channel to the Bay of Bengal. An estimated 100,000 fisherfolk depend on this ecosystem for their livelihood. It is here that the large industrial house of the Tatas, in partnership with the state government, has proposed an 'Integrated Shrimp Farming Project' to augment productivity and exports. However, local fisherfolk anticipate a host of problems with the coming of the project. Aside from the declining availability of fish for them – which will be the first consequence of omnivore entry into Chilika – they argue that the construction of large embankments, which the project demands, will increase threats from floods and water-logging. The project will also pollute the ecosystem with its artificial protein feed, and keep away the great flocks of migratory birds that now visit the lake. A social movement, in which students have joined hands with ecosystem people, has arisen to try to stop the project. Aside from numerous petitions and a press campaign, the movement mobilized 8,000 supporters in a demonstration in the state capital, Bhubaneshwar, in September 1991 (Dogra 1992). This is a classic conflict between omnivores and ecosystem people, with the former using the state to capture resources previously under the control of the latter.

FIGHTS IN THE FOREST

India's tropical soils are by and large poor in minerals and their cultivation demands substantial inputs of nutrients. The Indian cultivator has traditionally obtained these nutrients from non-cultivated lands surrounding the villages, either through loppings from trees, or by employing the livestock to convert grass into dung, which is then used as manure. Forest lands as grazing lands have therefore been an integral part of the subsistence base of India's ecosystem people. Not surprisingly, forest conflicts have been endemic ever since the British laid claim to India's vast forest tracts and attempted to wean the people away from their traditional use of these common lands.

Indeed, the origins of the Indian environmental movement can be fairly ascribed to that most celebrated of forest conflicts, the Chipko movement of the central Himalaya. Governmental officials, in their deferential way, tend to locate the origins of environmental concern in India to the love for nature frequently expressed by the long-time Prime Minister, Mrs Indira Gandhi. Yet as a 'grassroots' perspective the Chipko movement has an authenticity lacking in the interventions of politicians and scientists. As a powerful statement against the violation of customary rights to the forest by commercial timber operations, Chipko brought into sharp focus a wide range of issues concerning the country's forest policy which also impinged on the environment debate as a whole. (For a full account of the history and sociology of Chipko, see Guha 1989a.)

Owing to its novel techniques and Gandhian associations, the Chipko movement has rapidly acquired fame. Yet it is representative of a far wider spectrum of forest-based conflicts. In the tribal areas of central India, economic dependence on the forests is possibly even more acute than in the Himalayan foothills where Chipko originated. Here the 1970s witnessed escalating conflict between villagers and the forest administration in tribal districts of the states of Bihar, Orissa, Madhya Pradesh, Maharashtra and Andhra Pradesh. In tribal India, moreover, forest conflicts often have a sharper political edge. Thus in Bihar, they have been an integral element in the popular movement for a tribal homeland, while in the four other states mentioned above, the question of tribal forest rights has been actively taken up by revolutionary Maoist groups (see Sengupta 1982; Peoples Union for Democratic Rights 1982; Calman 1985).

Scholarly research inspired by the forest conflicts of the 1970s also revealed their long lineage. Indeed, local opposition to commercial forestry dates from the earliest days of state intervention. Before the inception of the Indian Forest Department in 1864, there was, by and large, little state intervention in the management of forest areas, which were left in the control of local communities. The takeover of large areas of forest by the colonial state constituted a fundamental political, social and ecological watershed: a political watershed, in that it represented an enormous expansion of the powers of the state, and a corresponding diminution of the rights of ecosystem people; a social watershed, in that by curbing local access it radically altered traditional patterns of resource use; and an ecological watershed, in so far as the emergence of timber as an important commodity was to fundamentally alter forest ecology (Gadgil and Guha 1992: Chs 5 and 6).

The imperatives of colonial forestry were largely commercial. From our point of view, its most significant consequence has been the intensification of social conflict between the state and its subjects. Almost everywhere, and for long periods of time, the takeover of the forest was bitterly resisted by local populations, for whom it represented an unacceptable infringement of their traditional rights of access and use. Hunter-gatherers, shifting cultivators, peasants, pastoral nomads, artisans – for all these social groups free access to forest produce was vital for economic survival, and they protested in various ways at the imposition of state control. Apart from forest laws, new restrictions on *shikar* for local populations (while on the other hand allowing freer hunting for sport by the British and the Indian elite) were another contributory factor in fuelling social conflict (Rangarajan 1992).

Throughout the colonial period, popular resistance to state forestry was remarkably sustained and widespread. In 1913, a government committee in the Madras Presidency was struck by the hostility towards the Forest Department, which was the most reviled government agency along with the Salt Department (likewise concerned with a commodity ostensibly low in value but of inestimable worth to every village household). Two thousand miles to the north, in the Garhwal Himalaya, a British official wrote at almost the same time that 'forest administration consists for the most part in a running fight with the villagers'. Popular resistance to state forestry both embraced forms of protest that minimized the element of confrontation with authority – e.g. covert breaches of the forest law – and organized rebellions that challenged the right of the state to own and manage forest areas. (For a comprehensive analysis of forest-based movements in modern India and a guide to sources see Gadgil and Guha (1992).)

Ironically, the post-independence period only witnessed an acceleration of this process. Economic development implied more intensive resource use which, in the prevailing technological and institutional framework, led inevitably to widespread environmental degradation. In the forestry sector, the industrial orientation became more marked, exemplified in the massive monocultural plantations begun in the early 1960s. Simultaneously, other development projects such as dams and mines exerted a largely negative influence on the forests.

Not surprisingly, the conflicts between the state and its citizens have persisted, and the Forest Department continues to be a largely unwelcome presence in the Indian countryside. However, forest conflicts in independent India have differed in one important respect from conflicts in the colonial period. Earlier, these conflicts emerged out of the contending claims of state and people over a relatively abundant resource; now these conflicts are played out against the backdrop of a rapidly dwindling forest resource base. In other words, a newer *ecological* dimension has been added to the moral, political, and economic dimensions of social conflicts over forests and wildlife.

Cumulatively, these processes have worked to further marginalize poor peasants and tribals, the social groups most heavily dependent on forest resources for their subsistence and survival. A long-time student of Indian tribals poignantly captures their frustration with state forestry thus:

> The reservation of vast tracts of forests, inevitable as it was, was . . . a very serious blow to the tribesman. He was forbidden to practise his traditional methods of [swidden] cultivation. He was ordered to remain in one village and not to wander from place to place. When he had cattle he was kept in a state of continual anxiety for fear they would stray over the boundary and render him liable to what were for him heavy fines. If he was a forest villager he became liable at any moment to be called to work for the forest department. If he lived elsewhere he was forced to obtain a license for almost every kind of forest produce. At every turn the forestry laws cut across his life, limiting, frustrating, destroying his self-confidence. During the year 1933–34 there were 27,000 forest offences registered in the Central Provinces and Berar and probably ten times as many unwhipped of justice. It is obvious that so great a number of offences would not occur unless the forest regulations ran counter

to the fundamental needs and sentiments of the tribesmen. A Forest Officer once said to me: 'Our laws are of such a kind that every villager breaks one forest law every day of his life.'

(Elwin 1964: 115)

Popular movements in defence of forest rights have raised two central questions regarding the direction of forest management. First, they have contended that the control of woodland must revert to communal hands, with the state gradually withdrawing from ownership and management. Second, the opposition to forest management has contrasted the subsistence orientation of villagers with the commercial orientation of the state. This latter contrast can be illustrated by two strikingly similar incidents, separated in time by a few months but occurring 3,000 km apart. The first took place in Kusnur village in the Dharwad district of the southern state of Karnataka. Protesting against the allotment by the state of village pasture land to a polyfibre industry which intended to grow eucalyptus on it, on 14 November 1987 the peasants of Kusnur and surrounding villages organized a 'Pluck-and-Plant' *satyagraha* wherein they symbolically plucked a hundred eucalyptus saplings and replaced them with useful local species. Less than a year later, and probably without knowledge of the Kusnur precedent, Chipko activists in the northern state of Himachal Pradesh were arrested on charges of causing 'damage to public property'. Their 'crime' had been to lead villagers in uprooting 7,000 eucalyptus saplings from a forest nursery in Chamba district to plant indigenous broad-leaved species in their stead. The Dharwad and Chamba episodes vividly illustrate the continuing cleavages between village interests and the commercial bias of state forestry (Kanvalli 1991; Modi 1988).

The clash between subsistence agriculturists and industry over the usufruct of state forests is only the most visible of forest conflicts. Localized opposition has also arisen among village artisans facing increasing difficulty in obtaining raw material from forest lands. Characteristically, the state has diverted to industrial enterprises resources earlier used for generations by artisans. But reed workers in Kerala, bamboo workers in Karnataka and rope makers using wild grass in the Siwalik hills of Uttar Pradesh have all resisted the Forest Department's desire to give preferential treatment to the paper industry in the supply of biomass from state lands.

However, in most areas forest-dependent artisans are yet to be politically organized. That is no longer the case with millions of tribals in central India for whom the collection and sale of 'minor' (i.e. non-wood) forest produce is vital to survival. For decades, tribals collecting non-wood forest produce have been severely exploited by merchants who control the trade. For these merchants, the most lucrative of all 'minor' forest produce is the tendu leaf, used in making the *bidi* or Indian cheroot. Over the past two decades, social activists have organized tendu leaf pluckers in a bid to increase their collection wages. On the eve of the 1991 plucking season, twenty-four organizations working among tribals in five contiguous states of central India unilaterally announced that they had fixed the price of tendu leaves at Rs 50 per 5,000 leaves (the merchants' acquisition rates varied from Rs 9 per 5,000 leaves in Bihar to Rs 25 in Maharashtra). In several areas, tribal forest labourers have been organized by left-wing revolutionaries, causing the alarmed traders to seek the protection of the state. Sadly but inevitably, violence has escalated in the tribal forest districts of Orissa, Madhya Pradesh, Maharashtra and Orissa (*The Statesman*, 17 March 1991; Peoples Union for Civil Liberties 1985).

MINES AND MISERY

Conflicts over fish and forest resources have both arisen out of the competing claims of omnivores and ecosystem people, each coveting the same resource but for different reasons. By contrast, the conflicts we now highlight are a consequence of the negative externalities imposed by one kind of economic activity – open cast mining – upon another – subsistence agriculture.

The most celebrated of mining conflicts took place in the Doon valley in northwest India. Home to the Indian Military Academy as well as the country's most famous public school, this beautiful valley is a favourite watering hole of the Indian elite. Here, the intensification of limestone mining since 1947 has led to considerable environmental degradation: deforestation, drying up of water sources and the laying waste through erosion and debris of previously cultivated fields. Opposition to limestone quarrying has come from two distinct sources. On the one side, retired officials and executives formed the 'Friends of the Doon' and the 'Save Mussoorie' committees to safeguard the habitat of the valley. They were joined by hotel owners in Mussoorie, worried about the impact of environmental degradation on the tourist inflow into this well-known hill station. These groups may fairly be characterized as NIMBY (Not In My Backyard) environmentalists, preoccupied above all with protecting a privileged landscape from overcrowding and defacement. On the other side, villagers more directly affected by mining were organized by local activists, many of whom had cut their teeth in the Chipko movement. Whereas the first group lobbied hard with politicians and senior bureaucrats, the latter resorted to sit-ins to stop quarrying. Finally, both wings collaborated in a public-interest litigation that resulted in a landmark judgment of the Supreme Court, which recommended the closure of all but six limestone mines in the Doon Valley (Dogra *et al.* 1983; Bandyopadhyay 1989).

At the height of the Dehradun limestone controversy, one of the valley's NIMBY environmentalists called – with characteristic disregard for the inhabitants of those areas – for mining to be shifted to the interior hills so that Dehradun and Mussoorie would be spared (Dalal 1983). Unknown to the writer, mining had already proceeded apace in the inner hills. Expectedly, it has met with resistance. In the Almora and Pithoragarh districts of Kumaun, soapstone and magnesite mining, by taking over or leading to the degradation of common forest and pasture land, has greatly reduced local access to fuel, fodder and water. With the onset of the monsoon, the debris accumulated through mining descends on the fields of adjacent villages. Meanwhile, tangible benefits to the village economy are few – and certainly inadequate in offsetting the losses due to declining agricultural productivity and biomass availability – with mining lessees preferring to bring in outside labour to act as a buffer between management and villagers.

Kumaun has a long heritage of social movements (Pathak 1987; Guha 1989a) and this has been invoked in the continuing struggles against unregulated mining. Social activists have worked hard to form village-level *sangarsh samitis* (struggle committees) in the affected areas. The Laxmi *ashram* in Kausani, started by Gandhi's disciple Sarla Devi, has been quite successful in involving women in these movements. In other instances, villagers have acted independently to protest against the damage done by open cast mining. The forms of struggle have been varied: *padayatras* (walking tours), *dharnas*, fasts and efforts to persuade miners to go on strike. In many of these protests, women – whose own domain is most adversely hit by mining – have played a leading role. Several mines have been forced to close down, and villagers have then turned their energies to land reclamation through afforestation (Institute of Social Studies Trust 1991; Joshi 1983a, b).

Another movement with broadly similar contours has taken place against bauxite mining in the southeastern state of Orissa. In the Gandhamardan hills of the Sambalpur district of the state, the public-sector Bharat Aluminium Company (BALCO) has been granted permission to mine a heavily forested area of about 900 ha. The foundation stone for the project was laid by the Chief Minister of Orissa in May 1983, and mining commenced two years later; but by the end of 1986, BALCO operations had been forced to a halt.

As in the Himalaya, bauxite extraction in Gandhamardan led quickly to deforestation, erosion and the pollution of water sources. Blasting operations were also perceived as a threat to the region's ancient temples, visited by pilgrims from long distances. Characteristically, protest originated in a series of petitions sent to senior officials and politicians. When these had no effect, students and social activists began forming village committees. A three-day *dharna* in front of the block development office in October 1985 was followed two months later by a blockade that prevented BALCO vehicles from proceeding up the Gandhamardan plateau to the mines. Private vehicles carrying materials for BALCO operations were also blocked and unloaded. In January and February 1986, the movement shifted to the site of BALCO's proposed railway line, close to the Orissa–Madhya Pradesh border. According to figures collected by a civil liberties group that visited the area, altogether 987 people were jailed in the course of the year-long struggle, including 479 women and 51 children (Peoples Union for Democratic Rights 1986; Concerned Scholars 1986).

Plate 17 This site of an abandoned iron mine in Karnataka's Western Ghats once supported primeval rain forest

Plate 18 South Indian granite is being exported in large volumes to Europe. Unregulated quarrying of granite is a significant cause of deforestation and disruption of watercourses

THE WILD AND THE SACRED

The conflicts considered so far have related to economic interests of different actors in the use of natural resources. There is yet another category, where aesthetic, recreational, scientific and even religious interests of one group clash with the economic interests of another group. There is for instance the opposition to the submersion by the Upper Krishna Reservoir of Kudala Sangama, the site famous as the place where the Kannada saint Basavanna is believed to have attained salvation. But the more striking examples of such conflicts, considered below, relate to nature conservation.

Ecosystem people all over the world have viewed themselves as members of a community of beings, in coexistence with fellow creatures be they trees, birds, streams or rocks. Many of these are revered and protected as sacred objects. Sacred is the Himalayan peak of Nanda Devi, as too is Talakaveri from where originates the River Kaveri in the Western Ghats. Entire patches of forests, or pools in river courses, or ponds may be treated as sacred and accorded protection against human exploitation. All individuals of a species like the rhesus monkey might be treated as sacred objects not to be harmed. Such age-old traditions of nature conservation have played an important role in conserving India's heritage of biodiversity.

Sacred groves are an element of this system, older even than Gautama Buddha, who was born in a sal forest sacred to the goddess Lumbini. A rich network of sacred groves once covered all India, and Dietrich Brandis, the first Inspector-General of Forests, was already lamenting the destruction of much of this network under the British system of forest management by the 1880s (Brandis 1884). But remnants of this network still exist and are today the very last representatives of the climax vegetation in areas such as the Maharashtra Western Ghats. In 1972, one of us became interested in inventorying these sacred groves as exemplifying folk conservation practices of Maharashtra. When this survey was being conducted a letter arrived from a village called Gani in the Shrivardhan *taluk* of coastal Maharashtra. The letter stated that the village had a well-preserved sacred grove, of 10 ha, which had been marked for felling by the Forest Department, and asked for our help in saving it. We visited this remote village, after trekking for some three hours across totally devastated hills in a region that receives over 3,500 mm of rain, to find the sacred grove, which was one of the last remaining pieces of greenery. The villagers said that the only perennial stream near their village flowed out of this grove, and they desperately wanted the authorities not to cut it down. Nobody, however, cared for their protests so that when they learnt from a sympathetic forest ranger that we were circulating questionnaires about sacred groves, they hoped we could help. And indeed we could, with the Forest Department agreeing to spare this grove, although one of the senior officials wondered why we were bothered about saving these stands of 'overmature timber'. Saving this grove was, however, an isolated success, and all over India the last of the sacred groves are being felled, often against protests by the local population (Gadgil and Vartak 1975; Gadgil and Subash Chandran 1992).

While the protection of sacred groves is a traditional folk practice of ecosystem people, the omnivores have their own, primarily recreational, interest in nature conservation. Thus some 4 per cent of India's landmass has now been put under some form of protected area, as wildlife sanctuary, national park or biosphere reserve. The guiding philosophy behind the management of these protected areas has been one of keeping the local people out by force of arms, on the theory that they are the sole cause of the degradation of nature. There is little real evidence for this theory; indeed, there is accumulating evidence that pressures from outside omnivores, not those of local ecosystem people, are the main problem. Despite this evidence, attempts to deprive local people of access to resources of protected areas continue, and are leading to sharp conflicts. These conflicts have been reported from all over the country, from the Rajaji National Park in the north to the Nilgiri Biosphere Reserve in the south, and from the Betla Tiger Reserve in the east to the Nal Sarovar Sanctuary in the west. Sometimes they can have tragically destructive effects, as when a hostile local population is moved to burn large areas of a National Park which they perceive to be against their interests – acts of arson which they have resorted to in two important national parks, Kanha in Madhya Pradesh (in 1989) and Nagarhole in Karnataka (in 1992). Ironically, where ecosystem people living in and around parks have often been denied their traditional rights to forest produce, within the reserves commercial exploitation has continued with the connivance of officials. An enquiry by the Peoples Union for Democratic Rights into tribal discontent with the management of the Simlipal Tiger Reserve in Orissa came to this stark conclusion:

> Thus in Simlipal the choice is not between no poaching and poaching but between . . . poaching by tourists and organized smugglers and occasional tribal poaching. The choice is not betwen no cultivation and cultivation but between large scale illegal denudation of forests and cultivation by tribals . . . The choice is not between [a] complete removal of human settlement and deforestation by tribals but between organized deforestation with the connivance of state agencies and limited deforestation by tribals. In the end the choice is not between an ecosystem without human interference and that with human interference but it is between interference by tribals and interference by smugglers, traders and pleasure-seekers. It is a choice between two sets of human beings.
>
> (PUDR 1986)

– a choice, in our terms, between omnivores and ecosystem people.

A well-documented case of such a conflict is located at the Keoladeo National Park at Bharatpur in Rajasthan. Keoladeo is an artificially created

Plate 19 Women carrying fuelwood out of the Ranebennur wildlife sanctuary in Karnataka. There is perennial conflict between local communities and official guardians of protected areas

wetland, a shallow body of water of over 450 ha created by bunds constructed in the eighteenth century. This wetland attracts tens of thousands of wintering waterfowl as well as supporting an enormous number of herons, egrets, storks and ibises that breed during the monsoon. It was once a site of enormous shoots – tens of thousands of ducks and teals in a day – by British aristocrats coming down as guests of the maharaja of Bharatpur. The area had also always served as a grazing ground for cattle and buffaloes from local villages and as a source of irrigation in the post-monsoon period.

After independence the area was constituted as a national park, and taken over from the Bharatpur maharaja, who yielded it with great reluctance. Then began to operate the omnivore theory of the necessity of keeping ecosystem people at bay. Not only forest officials, but scientists, both Indian and American, claimed that a ban on grazing would be highly desirable – in the total absence of any evidence on this score. Such a ban was finally imposed in the early 1980s, but without any alternative provision for fodder supply. Peasants protested, there was police firing, killing some of them, and the ban was forcefully implemented. The results have been disastrous for Keoladeo as a bird habitat. In the absence of buffalo grazing, *Paspalum* grass has overgrown, choking out the shallow bodies of water, rendering this a far worse habitat, especially for wintering geese, ducks and teals, than it ever had been before (Vijayan 1987).

Wild animals protected in the conservation area, or protected species, outside it, can often affect the people adversely in various ways. Thus wild boar protected by the Bhimashankar Sanctuary in Pune district every year destroy extensive areas of standing crops in tribal hamlets which adjoin the reserve. In 1987 and 1993, a voluntary organization working with the tribals attempted to quantify the extent of crop damage by wild boars. In twenty-five hamlets surveyed in 1987, about 96,000 kg of grain had been destroyed by wild animals – a total monetary loss to the farmers of Rs 232,000. Six years later, the damage was computed at 90,820 kilograms, valued at approximately Rs 453,000 at current prices. The average loss per person was estimated at Rs 800 per year – in the circumstances, a considerable sum of money. To this must be added the difficulties faced by the tribals as a consequence of restricted access to forest produce, more freely available to them before the constitution of the sanctuary (Anil Kapur and Kusum Karnik, pers. comm.).

These enormous losses suffered by ecosystem people in the interior rarely attract the attention of wildlife lovers in the city. Ironically, these are the people whose energetic lobbying of government has contributed to the rapid and somewhat unplanned expansion of the protected area network. Curiously, urban wildlifers are prone to be more sensitive to animal depredations in their own vicinity. Thus bonnet macaque troupes still living in parts of the city of Bangalore damage papayas and guavas, roses and marigolds in the gardens of houses. We can recall an elite conservation group talking of the need to relocate these monkeys out of Bangalore, and simultaneously of the need to educate peasants to agree to bear crop damage caused by elephants in hilly regions to the southwest of the city. Not only do farmers in this area suffer extensive crop damage, but around twenty-five people are trampled to death every year by elephants. As is the case all over India, there is very inadequate provision for compensating for crop damage, while the family of the person killed has the solace of a princely sum of Rs 5,000 obtained after much difficulty and red tape. Farmers employ a variety of devices to keep elephants at bay, including killing them with country-made guns. If the elephant so killed has tusks, it can fetch a very handsome income, often equivalent to several years of earnings for a landless tribal or small farmer (Sukumar 1989, 1994). Thus elephant poaching coupled to sandalwood smuggling has become a thriving business, especially in the dry scrub to the southwest of Bangalore, where agriculture is quite unproductive and there are few other avenues of employment in forest areas. Gangsters have come up to organize this business – the most notorious of them is one Mr Veerappan. He and his gang have killed dozens of forest and police personnel, but for years they have remained at large. This is because while Veerappan has provided employment to the local people, government employees are viewed merely as aliens and exploiters. As a result, villagers provide no information on Veerappan's whereabouts, permitting him to escape capture (*Frontline*, 19 May 1995).

THE ECOLOGICAL BASIS OF ETHNIC CONFLICT

In the ongoing inequitable process of development in India, not only are ecosystem people and ecological refugees everywhere denied access to resources, but certain regions are systematically drained to support others. The regions so impoverished usually have concentrations of the weakest segments of India's population, the tribals; the regions prospering at their cost have concentrations of those who wield economic and political clout, in metropolitan centres like Bombay and Delhi. The largest concentrations of India's tribal populations occur in the northeastern hills and in the so-called Jharkhand tracts extending over Bihar, West Bengal, Orissa and Madhya Pradesh in central India. Both these tribal areas are rich in forest resources; the northeast also has oil and the central Indian tracts have coal and iron ore. The forests are controlled by the state government, minerals by the central government. Although these areas generate enormous revenues, almost all of it flows out, leaving behind poorly paid wage labour and environments devastated by logging and mining. These inequities have fuelled numerous local separatist movements, be they the armed insurgencies of the Nagas, Mizos or Bodos, the Assam students' agitation or the Jharkhand movement. These movements demand more control over local resources and a greater share in the profits from their use – a demand vehemently opposed by omnivores sitting in Patna or Delhi.

The most long-standing of these movements is the Jharkhand agitation in the Chotanagpur region of Bihar. In an area rich in forest and mineral resources, the massive expansion of industrial and development projects has gone hand in hand with the impoverishment of the region's predominantly tribal population. Beginning in the late 1930s, there has been a consistent demand for a separate political entity ('Jharkhand') where the tribals would have effective control over their resources and their destiny. Foremost among the causes that lie behind the call for Jharkhand are the alienation of land and forest resources, the uncontrolled influx of outsiders who usually monopolize jobs and positions of power, and the grave neglect of infrastructural development by the government, so much so that a massive hydroelectric project which takes away power for industries would not even provide a light to local villages. Episodically, these demands have fuelled militant upsurges where tribals have surrounded or attacked government offices, or organized boycotts and blockades. Over the years, by systematically co-opting its leaders and through repression, the government of India has been able to thwart the creation of Jharkhand. All the same, the movement has not died out, and at regular intervals gathers renewed force (Devalle 1992; Ghosh 1991; Sengupta 1982). At last, in September 1994, the government of India has agreed to the creation of a Hill Council, where Jharkhand would remain within the state of Bihar but have a modicum of autonomy. It remains to be seen whether this concession will satisfy popular aspirations and diminish conflict.

In several of these regions, extremists have formed armed groups that virtually run parallel governments. These groups collect taxes, contending that this revenue must be collected and used locally, and not be permitted to flow out of the region to serve outside omnivores. Thus the Naxalites in the Bastar district of Madhya Pradesh demand Rs 5,000 from each lorry laden with bamboo going out to the paper mills. In the Chandrapur district of Maharashtra they have prevented operation of liquor shops that carried on a roaring business and were an important source of government revenue as well. Such movements have also precipitated major setbacks in the official nature conservation effort. Thus Naxalites have encouraged displaced tribals to encroach into and set fire to the Kanha Tiger Reserve, and Bodo extremist groups have permitted poaching in the Manas Tiger Reserve.

The inequities in contemporary India relate not only to control over land, water, fish, forest or minerals, but also to access to education, jobs in the bureaucracy, and the process of political decision making. There are growing social conflicts focused on each one of these concerns. Conflicts grow primarily because the gulf between omnivores and the dispossessed is continually widening. A most acute struggle therefore rages over entry into the omnivore class, whether through higher education, especially technical education, or more directly through a job in the government. The most recent manifestation of this conflict is the struggle over caste-based reservation in professional colleges and in government and public-sector employment. The principle of such reservation in favour of scheduled castes and scheduled tribes has long been accepted. It was long ago extended to other 'backward' castes in the southern states as a fallout of the anti-*Brahman* movement. In 1990, the Prime Minister accepted the Mandal commission report which would extend its coverage to employment in all central government undertakings. There was a surge of violent protests in the cities of northern India, marked by macabre incidents of self-immolation by upper-caste students.

The nation-wide debate sparked off by the Mandal controversy, however, failed to address itself to the deeper causes of why such a wide gulf exists in the first place between the omnivores and the dispossessed. This gulf has little to do with the fact that most members of backward castes can never hope to gain positions of power in government. Rather, its deeper causes are to be found in a pattern of development that has left the bulk of this class of people deprived of control over land, water, forest, minerals, education, employment and decision making. After more than four and a half decades of independence 38 per cent of males and 66 per cent of females in India remain illiterate. Notably, this lack of literacy is concentrated among the predominantly lower-caste ecosystem people. When, therefore, industries come to areas inhabited by them, as with the steel mills of Durgapur and Rourkela, all the well-paying jobs go to outsiders. A small number of the locals get ill-paid, unskilled jobs, while the majority merely suffer from the pollution and deforestation that inevitably follow the establishment of such industrial enclaves.

This chapter has provided an overview of the multitude of resource-based conflicts in present-day India. While these conflicts are endemic everywhere, some are more visible than others. They attract little attention when they concern only ecosystem people in the hinterlands, as with the landless encroaching on grazing lands valued by small peasants, or when they arise among ecological refugees, as with fights at public taps in city slums. The media pay far more attention to conflicts that pit omnivores either against each other as in urban land scams, or against ecosystem people as in the Chipko or Narmada Bachao *andolans*. In the colonial mode focusing on subjugation, these conflicts were essentially about the rapid drain of natural resources; here forest-related conflicts occupied centre stage. The conflicts have waned as forests have been exhausted over much of the subcontinent, while other sources of forest raw material such as imported Malaysian timber have provided a substitute. In the post-independence mode of governance as patronage, conflicts are sharpest where there are rich possibilities of cornering state-sponsored subsidies, as with river valley projects. Equally acute are conflicts over gaining a foothold in the omnivore class, as with the issue of caste-based reservations. But acute or subdued, endemic or suddenly erupting, resource-based conflicts have fissured Indian society along many, many axes. Their resolution is evidently an urgent national task. We offer our own perspectives in Part II of this book. But first let us locate the Indian environmental movement in the context of these conflicts.

4

IDEOLOGIES OF ENVIRONMENTALISM

THE MAJOR STRANDS

One may define an environmental movement as organized social activity consciously directed towards promoting sustainable use of natural resources, halting environmental degradation or bringing about environmental restoration. Viewed in this light India has a wide diversity of environmental movements involving members of one or more of our three categories of omnivores, ecosystem people and ecological refugees. In this multiplicity of movements one may discern seven major strands. Two of these are exclusively focused on nature conservation, one on aesthetic, recreational or scientific grounds and the other on the basis of cultural or religious traditions. The wildlife conservation movement, largely attracting urban omnivores, represents the first strand; the Bishnoi peasants of Rajasthan assiduously protecting *khejadi* trees and blackbuck, chinkara, nilgai and peafowl around their villages, the second. A third strand focuses on efficiency of resource use from a technocratic perspective. This has prompted the establishment of land use boards and integrated watershed development programmes, manned and run by omnivores.

However, the dominant strands in the Indian environmental movement are those that focus on the question of equity. These have largely arisen out of conflicts between omnivores who have gained disproportionately from economic development and ecosystem people whose livelihoods have been seriously undermined through a combination of resource fluxes biased against them and a growing degradation of the environment. Such movements most often tend to involve a small group of socially conscious omnivores working with larger numbers of ecosystem people or ecological refugees.

We might call these movements 'the environmentalism of the poor' to distinguish them from the environmentalism born out of affluence that is such a visible presence in the advanced capitalist societies of the West (Martinez-Alier 1990). There are four broad strands within these movements. The first emphasizes the moral imperative of checking overuse and doing justice to the poor, and largely includes Gandhians. The second emphasizes the need to dismantle the unjust social order through struggle, and primarily attracts Marxists. The third and fourth strands emphasize reconstruction, employing technologies appropriate to the context and the times. These might arise either out of the concerns of scientists or, more significantly, through the revival of community-based management systems. The latter include spontaneous village-level efforts to protect and sustainably use local wood lots or ponds, or to pursue environmentally friendly agricultural practices.

ENVIRONMENTAL STRUGGLES

In analysing environmental movements, one may distinguish between their material, political and ideological expressions. The material context, discussed in preceding chapters, is provided by the wide-ranging shortages of, threats to and struggles over natural resources. Against this backdrop, the political expression of Indian environmentalism has been the organization by social action groups of the victims of environmental degradation. Even the urban well-to-do, increasingly subject to noise and air pollution, and deprived of exposure to nature, might be viewed as victims of environmental degradation, and their organization into societies like the World Wide Fund for Nature (WWF-India) is an environmental movement of sorts. Indeed, some Western scholars such as Thurow (1980) have contended that environmentalism is predominantly an interest of the upper middle class of the rich countries, and that poor countries and poor individuals are simply not interested in environmentalism. It is abundantly clear, however, that Indian environmental movements very much involve the poor and the disadvantaged victims of environmental degradation. In the rest of our discussion, therefore, we shall almost exclusively focus on environmental movements involving struggles of ecosystem people.

Environmental action groups working with such people have embarked upon three distinct, if interrelated, sets of initiatives. First, through a process of organization and struggle they have tried, with varying degrees of success, to prevent ecologically destructive economic practices. Second, they have promoted the environmental message through the skilful use of the media, and, more innovatively, via informal means such as walking tours and eco-development camps. Finally, social action groups have taken up programmes of environmental rehabilitation (e.g. afforestation and soil conservation), restoring degraded village ecosystems and thereby enhancing the quality of life of its inhabitants.

Although these myriad initiatives may be construed, in the broad sense, as being political in nature, they have been almost wholly undertaken by groups falling outside the sphere of formal party politics. Across the ideological spectrum of party politics in India – from the Bharatiya Janata Party on the right to the Communist Party of India (Marxist) on the left – the established parties, whose higher-level leadership has become an integral part of the omnivore establishment, have turned a blind eye to the continuing impoverishment of India's natural resource base, and the threat this poses to the lives and livelihoods of vulnerable populations. In the circumstances, it has been left primarily to social action groups not owing allegiance to any political party – what the political scientist Rajni Kothari (1984) has termed 'non-party political formations' – to focus public attention on the linkages between ecological degradation and rural poverty.

Through the process of struggle, the spreading of consciousness and constructive work, action groups in the environmental field have come to develop an incisive critique of the development process itself. Environmental activists, and intellectuals sympathetic to their work, have raised major questions about the orientation of economic planning in India, its inbuilt biases in favour of the commercial–industrial sector, and its neglect of ecological considerations. More hesitantly, they have tried to outline an alternative framework for development which they argue would be both ecologically sustainable and socially just. Although perspectives within the movement are themselves quite varied, in its totality this fostering of a public debate on development options constitutes the ideological expression of the environmental movement.

CREATING AWARENESS

In most of the conflicts over natural resources, collective protests – against the agencies of the state or against private firms – have been closely accompanied by coverage in the print media. Sometimes, leading environmental activists – Sunderlal Bahuguna and Baba Amte come immediately to mind – themselves write signed articles in newspapers drawing attention to the struggle they are engaged in. More frequently, though, sympathetic journalists write on these struggles and their wider implications. Since the mid-1970s there has been a virtual explosion of environmental writing in English- and Indian-language newspapers and magazines. Among the most notable of such publications have been the citizens' reports on the state of the Indian environment and the magazine *Down to Earth* published by the Delhi-based Centre for Science and Environment, and the books and magazines brought out by the Kerala Sastra Sahitya Parishat, Kerala's popular science movement. With radio and television controlled by the state, the print media have played an important role in reporting, interpreting and publicizing nature-based conflicts in modern India.

And yet, in understanding the spread of environmental consciousness, one must not underestimate oral means of communication. For example, to increase popular awareness of deforestation and pollution the Kerala Sastra Sahitya Parishat has performed plays and folk songs in all parts of Kerala. In the neighbouring state of Karnataka, themes of environmental abuse and renewal have figured in the traditional dance-drama of the west coast,

Yakshagana. An activity that combines discussion and practical action is the 'eco-development' camp, widely used by action groups to promote afforestation and other forms of environmental restoration (see Bhaskaran 1990).

But in the sphere of communication, too, the most innovative technique of the environmental movement recalls its acknowledged patron saint, Mahatma Gandhi. This is the *padayatra* or walking tour. Used by Gandhi to spread the message of communal harmony and by his disciple Vinoba Bhave to persuade landlords to donate land to the landless, the *padayatra* has been enthusiastically revived by environmental activists. The first environmental *padayatra* was in fact undertaken by one of Bhave's own disciples. This was the Kashmir to Kohima trans-Himalayan march, covering 4,000 km, accomplished by Sunderlal Bahuguna and his associates in 1982–3.

The most notable *padayatra* of this ilk was the Save the Western Ghats March of 1987–8. Following seven months of preparation involving over 150 voluntary organizations (from the states of Kerala, Tamil Nadu, Karnataka, Goa and Maharashtra), the march commenced on 1 November 1987 simultaneously from the two extremities of this 1,600 km long mountain chain – Kanyakumari in Tamil Nadu and Navapur in the Dhulia district of Maharashtra. Three months later, marchers from the north and south converged at Ponda in Goa for the meeting that marked the march's conclusion. By then they had collectively covered 4,000 km of hill terrain, making contact with over 600 villages *en route*. The predominantly urban marchers themselves came from a variety of backgrounds and age groups. Their aim was threefold: to study at first hand environmental degradation and its consequences for communities living along the Ghats; to try and activate local groups to play a watchdog role in preventing further ecological deterioration; and to canvass public opinion in general (Vijaypurkar 1988; Hiremath 1988).

One of the objectives of the Western Ghats march, in which it largely succeeded, was to draw attention to threatened mountain ecosystems other than the Himalaya, whose plight had hitherto dominated the Indian environment debate. As a haven of biological diversity (1,400 endemic species of flowering plants alone) and the source of many rivers, the Ghats are as crucial to the ecological stability of peninsular India as indeed the Himalaya are to the Indo-Gangetic plain. Notably, the Western Ghats march inspired *padayatras* across other vulnerable mountain systems. In the winter following the Western Ghats campaign, a 'Save the Nilgiris' march was organized. Covering villages in four *taluks*, this march culminated in a public meeting at the hill station of Ootacamund on Christmas Eve 1988 (*The Times of India*, 15 December 1988). Again, a 'Save the Sivaliks' march was undertaken across 200 km of the Sivalik range in Jammu and Kashmir the winter following the Western Ghats enterprise, while in early 1991 a fifty-day march was carried out through the Eastern Ghats of Andhra Pradesh and Orissa. The latter effort, termed the 'Vanya Prant Chaitanya Yatra' (forest areas awareness march), focused on the interconnections between environmental degradation and tribal poverty, as exemplified by deforestation, pollution, land alienation and displacement (see Saraf 1989; Vinayak 1990). Most recently, a group of social activists, predominantly Gandhian in orientation, organized a two-month-long 'Aravalli Chetna Yatra' in late 1993, traversing over 600 km on foot through a mountain chain that extends over the states of Gujarat, Rajasthan and Haryana apart from the capital city of Delhi. The marchers drew particular attention to the damage caused by illegal mining and logging in the Aravallis (Kishore Saint, pers. comm.).

Our final illustration of an environmental *padayatra* highlights not a region but a threatened resource – water. This was the Kanyakumari march, organized by the National Fisherfolk Forum in April 1989 under the slogan 'Protect waters, protect life!' As in the Western Ghats, two teams started independently – one in a fishing village in Bengal on the east coast, the other near Bombay on the west coast. Making their way on foot and by van, the marchers organized a variety of meetings and seminars in villages along the way. Although initiated by organizations working among fisherfolk, the march had a wider ambit. Apart from declining fish yields, the marchers studied the pollution of coastal waters by industry and urban sewage, and the destruction of key ecosystems such as mangrove swamps and estuaries.

The objectives of the march, as enumerated by its organizers, were:

(a) To widen people's awareness of the link between water and life and to encourage popular initiatives to protect water;
(b) to form a network of all those concerned with these issues;
(c) to pressurize the government in evolving a sustainable water utilization policy, and to democratize and strengthen the existing water management agencies;
(d) to assess the damage already done, identify problem areas for detailed study, and evolve practices for rejuvenating water resources; and
(e) to revive and propagate traditional water conservation practices and regenerative fishing technologies.

(National Fisherfolk Forum 1989)

The marchers on both coasts converged in Kanyakumari, on the southernmost tip of India, on May Day 1989 (this culminating date reflecting the trade union connections of the organizers). An exhibition on water pollution and conservation, held at a local high school, was followed by a march to the sea. Here the participants, led by 100 women, took a pledge to 'protect waters, protect life'. Finally, a crowd of nearly 10,000, at least half of whom were women, wound their way to the public meeting that was to mark the culmination of the march. Sadly, an incident provoked by a government bus disrupting the marchers led to a police firing in which several people were killed, and the rally was called off. Despite its aborted ending, the Kanyakumari march had fulfilled its aim of highlighting the threats to a liquid resource which, in the Indian context, must be reckoned to be as important as oil (see Dietrich 1989; Kumar 1989).

As tactics of struggle and consciousness raising, the *satyagraha* and *padayatra* have received generous media coverage. Less visible, but equally significant, are the programmes of ecological restoration that various social action groups have undertaken. With the state's manifest inability to restore degraded ecosystems, many voluntary organizations – some exclusively involving local people, and others relying on outside catalysts – have taken it upon themselves to organize villagers in programmes of afforestation, soil and water conservation, and the adoption of environmentally sound technologies.

ENVIRONMENTAL REHABILITATION

In focusing on environmental rehabilitation in preference to struggle or publicity, some groups are merely reviving indigenous traditions of community control, while others have been variously influenced by the Gandhian tradition of constructive work, by religious reform movements or by the example of international relief organizations. Often, voluntary groups with a background of work in health care, education or women's uplift have turned in recent years to promoting sound natural resource management. Of a wide range of groups we have chosen here to highlight two initiatives involving contact with the outside world, and two others that exclusively involve local people.

We start with the Dasholi Gram Swarajya Mandal (DGSM), the group that pioneered the Chipko movement, under the leadership of Chandi Prasad Bhatt. One wing of Chipko, identified with Sunderlal Bahuguna, has preferred to connect Himalayan deforestation with national and global environmental concerns. The DGSM, however, has turned from struggle to reconstruction work at the grassroots. Over the past decade, the DGSM has concentrated chiefly on afforestation work in the villages of the upper Alakananda valley. Two notable features of these plantations have been the lead taken by women, and the high survival rate of saplings – an average of 75 per cent in contrast to the 14 per cent in Forest Department plantations. In addition, in heavily eroded landscapes volunteers have taken up appropriate soil conservation measures such as the plugging of gullies, the construction of small check-dams and the planting of fast-growing grass species. Finally, the DGSM has enthusiastically promoted energy-saving devices such as fuel-efficient cooking stoves and biogas plants (Centre for Science and Environment 1985; S.N. Prasad, pers. comm.; Mukul 1993).

A recent investigation by the Satellite Applications Centre at Ahmedabad underscores the efficacy of this approach. Cultivated lands constitute only 4 per cent of the total landscape in this mountainous terrain – 1 per cent less than land under permanent snow cover. By 1972, when the first satellite pictures came, over 9 per cent of the land, mostly close to roads, had come to be covered by landslides or degraded scrub. In the 1970s the efforts of DSGM focused on checking the pace of deforestation. The satellite imagery shows that despite these efforts another 2 per cent of the land came to be covered by degraded scrub and landslides between 1972 and 1982. But plantation efforts were beginning to pick up in this decade, and an equivalent amount of old wasteland was nursed back to tree cover in this period. The tide was fully turned around in the 1980s. In this period only 0.5 per cent of land was newly converted to wasteland. At the same time, over 6 per cent of land was successfully brought under newly planted tree cover between 1982 and 1992 (Space Applications Centre 1993).

Our next case study originated not in a movement but in a remarkable individual, Anna Saheb Hazare of the village of Ralegaon Shindi in the Ahmednagar district of Maharashtra. Ahmednagar is in a drought-prone region; speaking of the scarcity of water there, the *Bombay Chronicle* of 2 March 1913 called it 'the most unfortunate and heavily tried district in India'. Thus when Anna Hazare returned to the village on retirement from the army in the mid-1970s, food production was barely 30 per cent of its requirements. Quickly locating the problem as insufficient retention of rain-water, he organized villagers in building a series of storage ponds and embankments (*nallah bandhs*) along the low hills surrounding the village. Very soon, runoff was reduced and aquifers recharged, and the ground water table rose considerably. There is now sufficient water for household use and irrigation, and crop yields have increased dramatically (the village has even started exporting food). Alongside, Hazare has mobilized villagers to plant 400,000 saplings. With his village now acknowledged as a model of eco-restoration through self-help, Hazare is training volunteers to work in other villages. He has simultaneously launched a movement against corruption in government-run forestry and drinking water programmes. Awarded the Padma Shri, a high national honour, Hazare returned the award to the Government of India in April 1994, following its failure to take effective action against forest officials accused of corruption (Rai *et al.* 1991; *Indian Express*, 18 April 1994).

Chandi Prasad Bhatt and Anna Hazare are among India's most celebrated environmentalists. Their own exposure to the wider society, through the Sarvodaya movement in the one case and army employment in the other, undoubtedly helped crystallize the ideas and strategies of action which they then applied in their own localities. However, there are many other initiatives, often totally unknown to the outside world, in which a group of local people have spontaneously organized efforts at eco-restoration and the sustainable use of natural resources. We report here two examples, previously unrecorded in the literature, that we are personally acquainted with.

The first of these is from two villages, Hosdurga and Rampura, in the semi-arid *taluk* of Pavgada on the Karnataka–Andhra Pradesh border. In this undulating terrain, the hillocks were once wooded with many hardy species.

But gradually they have been shorn of all tree cover, often through the sale of timber by local villagers to charcoal merchants. Some ten years ago a group of forty youths of Hosdurga, who had organized a small mutual fund for their own purposes, decided to reforest the hillock near their village. They sought and obtained the co-operation of the representative political body, the village *panchayat*. Investing some of the money from their mutual fund, they employed a watchman and strictly protected the emerging regeneration on the hillock. The vegetation is now coming back, and the group makes a little money by permitting harvests at a moderate level. The group has subsequently assumed an active role in other village development activities as well. Witnessing this success, a group from Rampura, a hamlet of the village of Hosdurga, has similarly taken to protecting the forest on a hillock near their habitation.

Our last example comes from the state of Manipur on the India–Myanmar border. Hill areas of this state are inhabited by shifting cultivators who led almost completely isolated lives until about 1910. Their traditional system of shifting cultivation involved leaving intact substantial patches of forest areas, abodes of nature spirits where cutting was taboo. These tribals embraced Christianity between 1910 and 1960, and on conversion cut down almost all these sacred forests. The results were disastrous, with fire from plots being brought under cultivation entering the villages and reducing them to ash. In many of the now Christian villages, such as the Gangtes village of Saichang in the Churchandapur district of Manipur, villagers have re-established a so-called 'safety forest' fringing the entire habitation. This safety forest is given strict protection, including a ban on harvest of canes, which have a lucrative market. While no longer believed to be an abode of nature spirits, this safety forest receives community protection, including the traditional punishment to any offender of having to sacrifice a pig and give a feast to the entire village. These tribals have also re-established protected bamboo forests from which no shoots are harvested as food, while bamboos may be harvested only for use in the construction of one's own house.

As these examples show, reconstruction work may proceed hand in hand with struggle. Yet in many other instances, groups temperamentally unsuited to confrontation have done estimable work in promoting environmentally benign technologies and in rehabilitating degraded lands. All in all, reconstruction work constitutes a valuable third front of the environmental movement, complementing the activities of consciousness building and popular resistance to state policies.

GREEN DEVELOPMENT

Individual groups working in the environmental field are typically confined to a small area. In the past decade, some attempts have been made to develop a macro-level organization to co-ordinate these varied groups and activities.

Plate 20 A procession through Harsud, a town that was about to be submerged under the rising waters of the Narmada

This process received a considerable boost with the rally against 'destructive development' held in Harsud in September 1989. In a follow-up meeting held in Bhopal in December – to coincide with the fifth anniversary of the gas tragedy in that city – groups that participated in the Harsud rally initiated the formation of the Jan Vikas Andolan ('People's Development Movement') or JVA, a loosely knit national level organization to co-ordinate local struggles, chiefly of ecosystem people.

Over the past four years, the JVA has had meetings in different parts of the country, involving a wide range of movements and individuals. In defining itself as a movement against the existing pattern of development, the JVA's own objectives are fourfold:

(a) to co-ordinate collective action against environmentally destructive policies and practices;
(b) to provide national solidarity to these struggles;
(c) to mobilize wider public opinion on the need for a new development path; and
(d) to work towards an alternative vision, ecologically sustainable and socially just, for India's future.

(Jan Vikas Andolan 1990)

To this end, it has joined groups representing construction workers, fisherfolk and other non-party formations in a national alliance of people's movements (*The Times of India*, Sunday edition, 3 April 1994).

Social action in the three generic modes outlined above constitutes the bedrock of the Indian environmental movement. While such activism has characteristically been localized, with most groups working within one district, the links between the micro and macro spheres have been made most explicit (recent initiatives like the JVA and the national alliance excepted) through the environmentalists' critique of the ruling ideology of Indian democracy, that of imitative industrialization. Environmentalists have insistently claimed that the intensification of natural resources conflict is a direct consequence of the resource-intensive, capital-intensive pattern of economic development, modelled on the Western experience, followed since independence. The resource illiteracy of development planning, they claim, is directly responsible for the impoverishment of the resource base and of the millions of rural people who depend on it (see also Jan Vikas Andolan 1990).

While there is widespread agreement within the environmental movement as regards the failures of the present development model, there is little consensus on, indeed inadequate effort at working out, plausible alternatives. Here we might identify three distinct ideological perspectives within the movement. It is of course entirely possible that none of the ideologies so identified is present in a particular struggle, or that adherents of all three viewpoints might participate unitedly in a specific initiative. However, interaction over many years with groups spread all over India does suggest that the three strands analysed below are the dominant ideologies of Indian environmentalism.

CRUSADING GANDHIANS

The first strand, which we may call 'crusading Gandhian', relies heavily on a moral/religious viewpoint in its rejection of the modern way of life. Here, environmental degradation is viewed above all as a moral problem, its origins lying in the wider acceptance of the ideology of materialism and consumerism, which draws humans away from nature even as it encourages wasteful lifestyles. Crusading Gandhians argue that the essence of 'Eastern' cultures is their indifference, even hostility, to economic gain. Thus, if India were to abandon its pursuit of Western models of economic development, it would only be returning to its cultural roots. These environmentalists call, therefore, for a return to pre-colonial (and pre-capitalist) village society, which they uphold as the exemplar of social and ecological harmony. Gandhi's own invocation of Ram Rajya (the mythical but benign rule of King Rama) is here being taken literally, rather than metaphorically. In this regard crusading Gandhians frequently cite Hindu scriptures as exemplifying a 'traditional' reverence for nature and lifeforms.

Crusading Gandhians have worked hard in carrying their message of moral regeneration across the country and indeed across the globe. They have sharply attacked the stranglehold of modernist philosophies – particularly those upholding rationalism and economic growth – on the Indian intelligentsia. Through the written and spoken word, they propagate an alternative, non-modern philosophy whose roots lie in Indian tradition (for a statement by the leading crusading Gandhian, the Chipko activist Sunderlal Bahuguna, see Bahuguna (1983); for an argument by a feminist follower of Bahuguna see Shiva (1988); and for more sophisticated intellectual treatments in the same vein, see Nandy (1987 and 1989)).

ECOLOGICAL MARXISTS

The second trend, in many ways the polar opposite of the first, is Marxist in inspiration. Marxists see the problem in political and economic terms, arguing that it is unequal access to resources, rather than the question of values, which better explains the patterns and processes of environmental degradation in India. In this sharply stratified society, the rich destroy nature in the pursuit of profit, while the poor do so simply to survive (the crusading Gandhians would tend to deny altogether that the poor also contribute to environmental degradation). For ecological Marxists, therefore, the creation of an economically just society is a logical precondition of social and ecological harmony. In their practical emphasis, socialist activists concentrate on organizing the poor for collective action, working towards their larger goal of the redistribution of economic and political power.

While including various Naxalite and radical Christian groupings, in the Indian context ecological Marxists are perhaps most closely identified with the People's Science Movements (PSMs), whose initial concern with taking 'science to the people' has been widened to include environmental protection. Ecological Marxists can be distinguished from Gandhians in two significant respects: their unremitting hostility to tradition (and corresponding faith in modernity and modern science) and their relatively greater emphasis on confrontational movements (for representative statements of this viewpoint, see Kerala Sastra Sahitya Parishat 1984; Raghunandan 1987).

APPROPRIATE TECHNOLOGISTS

Crusading Gandhians and ecological Marxists can be seen as being, respectively, the 'ideological' and 'political' extremists of the Indian environmental movement. Because of their ideological purity and consistency, their arguments are often compelling, albeit to different sets of people. In between these two extremes, and occupying the vast middle ground, lies a third tendency, which may be termed (less controversially) as 'appropriate technology'. This strand of the environmental movement strives for a working synthesis of agriculture and industry, big and small units, and Western and Eastern (or modern and traditional) technological traditions. Both in its ambivalence about religion and in its criticism of traditional social hierarchies it is markedly influenced by Western socialism. Yet in its practical emphasis on constructive work, it taps another vein in the Gandhians' tradition. Thus appropriate technologists have done pioneering work in the generation and diffusion of resource-conserving, labour-intensive and socially liberating technologies. Their emphasis is not so much the Marxists' challenging of the 'system', or the Gandhians' the system's ideological underpinnings, as it is demonstrating in practice a set of socio-technical alternatives to the centralizing and environmentally degrading technologies presently in operation (see Reddy 1982; Agarwal 1986; Bhatt 1984).

All three tendencies are represented in that most celebrated of environmental initiatives, the Chipko movement (Guha 1989a). The Gandhian trend, associated above all with the figure of Sunderlal Bahuguna, is best known outside the Himalaya. The Marxist trend within Chipko has been represented by the Uttarakhand Sangarsh Vahini, a youth organization that has organized popular movements against commercial forestry, unregulated mining and the illegal liquor trade. Finally, the appropriate technologists are represented by the organization under whose auspices the movement began, the Dasholi Gram Swarajya Mandal, whose fine work in ecological restoration has already been alluded to.

These contrasting perspectives may be further clarified by examining each strand's attitudes towards equity and science, as well as their style and scale of activism. Most crusading Gandhians reject socialism as a Western concept: this leads some among them to gloss over inequalities in traditional Indian society, and yet others even to justify them. Clearly the Marxists have been most forthright in their denunciations of inequality across the triple axes of class, caste and gender. The appropriate technologists have been sufficiently influenced by Marxism to acknowledge the presence and pervasiveness of inequality, but have rarely shown the will to challenge social hierarchies in practice. Attitudes towards modern science and technology also vary widely. The Gandhians consider science to be a brick in the edifice of industrial society, and responsible for some of its worst excesses. Marxists yield to no one in their admiration, even worship, of modern science and technology, viewing science and the 'scientific temper' as an indispensable ally in the construction of a new social order. Here, the appropriate technologists are the most judicious, calling for a pragmatic reconciliation between modern and traditional knowledge and technique, to fulfil the needs of social equity, local self-reliance and environmental sustainability.

Appropriate technologists prefer to work on a micro scale, a group of contiguous villages at best, in demonstrating the viability of an alternative model of economic development. The Gandhians have the largest attempted reach, carrying their crusade on world-wide lecture tours. They have often tended to think globally and act globally, even as the appropriate technologists have acted locally and occasionally thought locally too. The Marxist groupings work in the intermediate range, at the level of a district perhaps, or (as in the case of the Kerala Sastra Sahitya Parishat) the level of a state. Finally, the three strands also differ in their preferred sectors of activism. Their rural romanticism has led the Gandhians to emphasize agrarian environmental problems exclusively , a preference reinforced by their well-known hostility to modern industry. While appropriate technologists do recognize that some degree of industrialization (though not of the present resource-intensive kind) is inevitable, in practice they too have worked largely on technologies aimed at relieving the drudgery of work in the village. Here it is the ecological Marxists, with their natural constituency among miners and workers, who have been most alert to questions of industrial pollution and workplace safety.

Crusading Gandhians, appropriate technologists and ecological Marxists represent the three most forceful strands in the Indian environmental debate. But two other points of view also have a significant measure of support, especially among the omnivores. First, we have the Indian variant of that vibrant strand in global environmentalism, the wilderness movement. Indian naturalists have provided massive documentation of the decline of natural forests and their plant and animal species, urging the government to take remedial action (see Krishnan 1975). Although their earlier efforts were directed almost exclusively towards the protection of large mammals, more recently wildlife preservationists have used the scientific rhetoric of biological diversity and the moral arguments in favour of 'species equality' in pursuit of a more extensive system of parks and sanctuaries and a total ban on human activity in protected areas (Guha 1989b).

SCIENTIFIC CONSERVATION

We come, finally, to an influential strand of thinking within the state and state agencies. Focusing on efficiency, this strand might be termed 'scientific conservation' (Hays 1957). Pre-eminent here is the work of B.B. Vohra, a senior bureaucrat who was one of the first to draw public attention to land and water degradation. In a pioneering and impressively thorough paper (Vohra 1973), he documented the extent of erosion, waterlogging and other forms of land degradation. There was, he noted, no country-wide organization or policy to deal with these problems; nor was there co-ordination between concerned government departments. For Vohra, as for the early scientific conservationists (see Hays 1957), the solution lies in the creation of new ministries and departments to deal with problems of environmental degradation. The central government, he has written, 'has no option but to obtain a commanding position for itself in the field of land and soil

Table 2 Ideological preferences of the various strands of the Indian environmental movement

	Crusading Gandhians	*Ecological Marxists*	*Appropriate Technologists*	*Scientific Conservation*	*Wilderness Enthusiasts*
Polity	Highly decentralized democracy, 'village republics'	Dictatorship of the proletariat	Decentralized democracy, with women, low-caste participation	No firm view	No firm view
Decision making	Highly dispersed power of decision making	Centralized planning	Decentralized planning	Centralized planning	Strongly centralized administration
Society	No firm view	Economically equitable, but centralized political power	Economic and political equity	No firm view	No firm view
Economy	Mixed economy	State occupying 'commanding heights'	Mixed economy	Mixed economy	No firm view
Scale of economic enterprises	Predominantly small, village level	Predominantly large	Focus on small, complemented by large	No firm view	No firm view
Appetite for consumption	Limited through moral choice	Limited only on grounds of equity	Limited on grounds of both equity and ecology	Unlimited	Unlimited
Linkages to global economy	Weak	Weak	Weak	Weak	No firm view
Rate of technological change	Exceedingly low	High	Moderate	No firm view	No firm view
Commitment to military expenditure	Very weak	Strong	Weak	No firm view	No firm view

management through financial and administrative measures' (Vohra 1973: A12; but see also Vohra 1980, 1982).

Neither wilderness protection nor scientific conservation commands a popular following, yet each has had a considerable influence on government policy. Both tendencies look upon the state as the ultimate guarantor of environmental protection, and their energetic lobbying has informed stringent legislation in pursuance of this ideal – as for example the Wildlife Protection Act of 1972 (modified in 1991), the Forest Conservation Act of 1980, and the Environment Protection Act of 1986. However, in so far as neither group is cognizant of the social roots of environmental use and abuse, they tend to be dismissed as 'elite' conservationists by environmentalists owing allegiance to Gandhian or Marxist traditions.

So much for a thumbnail sketch of the main ideological strands of the Indian environmental movement. Table 2 summarizes their respective positions on a series of choices relevant to a new developmental paradigm. It is useful to construct such a table in order to bring out, first, that only Gandhians and Marxists have an overall, largely consistent philosophy of development, and, second, that there is very little agreement on any of the pertinent issues. Indeed, the ideological debate has been marked by a level of acrimony and abuse perhaps only to be expected in a youthful, radical movement – but distressing nevertheless. Little wonder then that the environmental movement has been quite unable to articulate a coherent alternative to correct the many shortcomings it has been so persistently fighting against. This has allowed the proponents of 'business as usual' (the ideologists of the omnivores) to dub environmentalists as being 'anti-progress', or even agents of foreign powers out to sabotage India's forward march. Such criticism must properly be met on its own ground, by articulating a coherent alternative path of development that accepts the fact that an overwhelming majority of human beings are engaged in the pursuit of their own self-interest. Part II of the book sketches out the elements of such an alternative. Chapter 5 sets out the broad philosophical underpinnings of the alternative paradigm. The following chapters then investigate three critical sectors, those of science and information, forestry, and population respectively. The conclusion highlights the resources we can draw upon in this effort.

Part II

THE INDIA THAT MIGHT BE

5

CONSERVATIVE-LIBERAL-SOCIALISM

A SYSTEM IN TROUBLE

India is a country full of contradictions. On getting up in the morning, we are expected to beg forgiveness from Mother Earth for stepping on her:

> Samudravasane prithvi, parvatastan mandale, Vishnupatni namastubhyam, padasparsham kshamaswa me!
>
> (O earth, consort of Vishnu, the Lord of creations, with mountains for thy breasts, and oceans for thy garments, forgive me for stepping on you.)

But not only do we not mind stepping on the earth, we blithely tolerate disasters like the Bhopal gas leak. India gave birth to Gautama Buddha, in a sacred grove of sal trees dedicated to the goddess Lumbini; Buddha achieved enlightenment under a peepul tree and preached a doctrine of compassion towards all creatures on earth. Today we are cutting down peepul and banyan trees, protected over centuries, to bake bricks to build our cities and to crate mangoes sent to the Middle East. We respect Mahatma Gandhi as the Father of the Nation; above all he wanted independent India to be rejuvenated as a land of village republics. But over the past forty-eight years, we have systematically sabotaged attempts to empower village people to control and manage their own destiny. We acknowledge a debt of gratitude to Jawaharlal Nehru, who wholeheartedly supported the development of modern science and technology in India in the belief that a scientific temper was the key to the rational development of the country's resource potential. But we continue to import almost all the technology we need, while using the prestige of science to push through development programmes without the open, fearless scrutiny that is so central to the scientific approach. B.R. Ambedkar, born into an untouchable caste, is called the father of the Indian constitution. But while we continue to extend caste-based reservation in educational institutions and government jobs well beyond the time limit Ambedkar himself had set, masses of our lower-caste peasants remain assetless in the absence of genuine land reform over most of the country. We have had Marxist parties in power in one or more states; one of these went out of its way to lure a private rayon mill by promising virtually free supply of forest raw material, and no checks or controls on polluting effluents.

For four decades after independence, India adopted a model of centralized socialist planning for all-round development. Yet no integrated view of the development process ever emerged. Planning degenerated into a process of merely allocating the state funds to meet various sectoral, often disparate, demands, depending on the relative clout of the interests being served. India, as the economic historian Dharma Kumar has observed (pers. comm.), is now a 'nation of grievance collectors' – the state is happy to submit to grievances in turn, even if they be contradictory. Thus we simultaneously have money sanctioned for giving loans to buy goats, and to undertake afforestation on the presumption that goats will be totally banned.

The previous chapters abundantly document the conflicts that have resulted from these manifold contradictions. The solutions sought for these problems too have been piecemeal, more in the nature of fire fighting than a systematic strategy tackling the root causes. This is as true of political problems of separatism and communalism as it is of environmental degradation.

Environmental activists have demanded that 'mega' projects like the Narmada or Tehri dams or the Gandhamardan mine be abandoned. But they have given little thought to how they can succeed in stopping them when the projects permit a narrow elite to corner large amounts of resources at state expense; and when the total lack of accountability in the way projects are planned and executed permits an alliance of politicians, bureaucrats and contractors to misappropriate a significant fraction of the resources deployed. There has been public agitation against pollution of water as in the Chazhiyar river case, or against air pollution as with the Bhopal gas leak. But can polluting industries really become motivated to mend their ways so long as their victims have no clout and when there is no open, publicly transparent process of the monitoring of pollution levels? There have been hesitant moves to create a stake for villagers in the health of nearby forest areas. But the activists who support these moves have not adequately examined the role of the high level of subsidies that forest-based industry enjoys, or the ways in which the forest bureaucracy can undermine popular participation. Wilderness lovers have lobbied for the total protection of some of the country's few remaining natural areas, without thinking clearly of how this is feasible when the iron triangle can profit so handsomely from exploiting these tracts, while the local tribals and peasants are being increasingly impoverished.

Some Indian environmentalists have gone so far as to contend that India must altogether eschew industrial development. But they have not squarely addressed the question of how this is possible in an increasingly unified world dominated by industrial countries, and where military might depends heavily on technological and industrial capabilities. More recently a demand was raised that India should decide not to pay its international debt and refuse to honour the conditions imposed by the World Bank for being given further loans. But again no thought has been given to how India can actually do this and still continue to function, given the foreign exchange needs of an economy that depends on large-scale import of petroleum products and a defence system which continually demands sophisticated hardware that India itself cannot manufacture.

Fighting against huge odds and the awesome power commanded by omnivores, the environmental movement in India has tended to be excessively defensive, taking on projects one by one. But surely it is time to go beyond the fire-fighting approach. We need to look at the process of development in a holistic fashion, while proposing a broad-based strategy for tackling the enormous difficulties we face as a nation. In our understanding, the system-wide difficulties arise out of six root causes:

(a) Ecosystem people have been suffering from an increasingly circumscribed access to natural capital, the resource base on which they still depend to fulfil many of their basic needs. This is because of the shrinkage of this resource base, as grazing lands are encroached or overgrazed, or natural forests give way to eucalyptus or *Acacia auriculiformis* plantations; and because the ever-growing state apparatus increasingly hinders them from using resources, as when 'open-access' revenue wastelands are taken over as strictly controlled reserved forest lands.

(b) Ecosystem people have very limited access to human-made capital, i.e. the resources of the organized industry–services sector. This is because employment in this sector has grown at a much slower pace than the population, and because education has failed to reach the large masses of ecosystem people, who must therefore eke out a living through unskilled labour on farms or in the informal sector.

(c) The process of building of human-made capital has been highly inefficient and greatly destructive of natural capital. This is because it has been conducted as a monopoly of a state apparatus without any public accountability, because the beneficiaries of these state-mediated interventions are given access to resources at highly subsidized rates and therefore do not care if the process is grossly inefficient, because the state apparatus has failed to force private enterprise to internalize environmental costs and finally because the cost of destruction of natural capital is passed on to the masses of people who are largely assetless, illiterate and, despite the democratic system, have no role in deciding on the direction in which the development process is moving.

(d) Omnivores are establishing, with the help of state power, an ever stronger hold over natural capital, as witness the displacement of Narmada refugees against their wishes, without any appropriate plans for their resettlement. Since the omnivores can pass on to others the costs of degradation of natural capital, their stranglehold promotes patterns of inefficient, non-sustainable use.

(e) Omnivores have a strong hold over human-made capital to the exclusion of both ecosystem people and ecological refugees. This means that the masses must subsist primarily through unskilled labour. In consequence they have no incentive to invest in quality of offspring, but instead produce large numbers of them, contributing to continuing population growth and adding to the resource crunch.

(f) There are large-scale outflows adversely affecting natural capital, whether this be iron and manganese mining silting up estuaries, overfishing in the sea or overgrazing to support the export of leather goods. These pressures are a result of the country's heavy dependence on import of technology and petroleum products, rooted in excessive concentration of economic development in a few islands of prosperity, as well as of high levels of demands for imported military hardware.

THE GANDHIAN WAY

Three among the existing political ideologies in India are, at first sight, comprehensive enough to address the whole range of issues pertinent to the development debate. These are the Gandhian, Marxist and liberal-capitalist philosophies. Gandhism, which is very much the dominant strand in India's environmental movement, is grounded above all in a moral imperative. With respect to the six issues raised above it proposes that:

(a) Ecosystem people should be given far greater access to and control over the natural resource base of their own localities. Ecosystem people should also be given an important role in a new, largely decentralized system of governance.

(b) Ecosystem people should remain content with their requirements of subsistence, without aspiring to greater access to material goods.

(c) The process of building up human-made capital at considerable cost to natural capital should be halted, by simply giving up the endeavour to step up resource use, to industrialize, or to intensify cultivation.

(d) and (e) Omnivores should not aspire to enhance their own material consumption, and in consequence give up their attempts to establish a stronger hold over the nation's natural and human-made capital.

(f) India should check the drain of its natural capital to the outside world by doing away with the need for foreign exchange through acceptance of a way of life with very low material demands, and a foreign policy based on non-violence leading to low military demands.

The serious flaw in the Gandhian philosophy is its emphasis on voluntary restraint on material consumption, and on the voluntary surrender of power.

While small numbers of people may do this, most people appear voluntarily to accept restraint on consumption or social power only if they are convinced that to do so is in their self-interest. Now restraint over resource consumption will be in the self-interest of an individual or a group of people only if such a group itself bears the cost of excessive levels of consumption. For instance, Manipur hill tribals have agreed among themselves not to harvest shoots or market bamboos from village forests because they are convinced that such restraints are essential to ensure an adequate supply of bamboo for their own requirements of house construction. The only route to a proper regime of restraint in resource use is therefore to pass on resource control to social groups who would themselves reap the benefits of prudent use. Today these are by and large India's ecosystem people; hence the Gandhian prescription of empowering them is quite sensible.

But the accompanying prescription that such a transfer of power should be a voluntary act on the part of the omnivores is obviously impractical. This, for instance, is the lesson of Vinoba Bhave's largely fruitless drive to bring about land reform through the voluntary surrender of land by larger land holders. For even where land was so gifted, Gandhian workers could never organize peasant recipients to put it to effective use. Indeed, it has been reported that when urged to experiment with new forms of social organization in villages where all lands had been surrendered, the Gandhians were afraid that such attempts would release forces that would have widespread violent repercussions and therefore withdrew from the scene. This and all other historical experience suggests that moral exhortations, Gandhian or otherwise, are unlikely to work widely.

India's omnivores are increasingly being swept into the global frenzy of consumption, which is daily fuelled by glossy magazines and satellite television. In the country's relatively open society the masses too are quite aware of and strongly attracted towards this consumer culture. While it appears materially impossible for India's 150 million families to own refrigerators and automobiles, they are all bound to aspire to do so. If meeting these aspirations is going to exact unacceptable environmental costs, then ways should be found of compelling all of India's citizens to share these costs and therefore accept restraints on consumption. But Gandhian moral exhortations are unlikely to bring this about.

Present patterns of resource consumption by India's omnivores cannot be sustained without heavy dependence on imports. Given the failure of India's industrial development to generate innovative products that can compete on the world market, these foreign exchange needs are met by draining the nation's natural resources. Again, Gandhian prescriptions seem incapable of getting us out of this bind. For what is required is a larger transformation of the system of resource use existing today.

Thus the key Gandhian prescriptions that make perfect sense are that ecosystem people must be empowered, and that material consumption should be maintained within limits compatible with a reasonably equitable sharing of the products of nature and the economy. However, the means suggested, that this should be brought about by voluntary acceptance by all people, in and out of power, as a moral imperative, seems impracticable.

THE MARXIST UTOPIA

While Gandhians view environmental problems as being caused primarily by materialistic greed, Marxists lay the blame at the door of capitalist exploitation simultaneously of nature and the working classes. By and large, they wholeheartedly approve of a concerted, nation-wide effort at stepping up natural resource use, so long as this is controlled and guided by a state acting on behalf of the people. With respect to the six key issues outlined above therefore, the Marxist approach may be summarized as being:

(a) Ecosystem people should be given far greater access to and control over the natural resource base. The leftist governments of West Bengal and Kerala indeed lead the country in efforts at land reform, involving local people in forest management, and in the establishment of decentralized political institutions at village and district levels.

(b) Ecosystem people should have better access to human-made capital. Again the left-oriented government of Kerala leads the country in taking literacy to all and in providing fuller access to employment in the modern industries–services sectors.

(c) Leftist state governments have, however, been no more successful in tackling the great waste and inefficiency in the process of conversion of natural into human-made capital. This problem arises in part from the naive Marxist faith in an all-powerful state apparatus which in practice behaves as irresponsibly (in an environmental sense) in India as it did in the erstwhile communist countries of eastern Europe.

(d) and (e) Indian Marxists, while in power, have indeed taken some steps to break the monopolistic access of omnivores to the capital of natural and human-made resources, but have not done enough to curb the state apparatus, which has become a significant component of the omnivore complex.

(f) Marxists wish to reduce the drain of natural resources to capitalist countries, and are active today in opposing what they view as the US-led conspiracy to lay India open to further exploitation through GATT and the series of economic reform measures demanded by the World Bank. But in the past they were happy enough with hefty exports of natural-resource-based goods such as tea and leather to the Soviet Union; nor have they any clear analysis of how to tackle India's compulsion to earn foreign exchange.

In fact, the emerging evidence of environmental degradation in eastern

Europe and the former Soviet Union abundantly demonstrates that a monolithic state apparatus behaves in a terribly wasteful and environmentally insensitive fashion. This is in part because most state employees have inadequate motivation to function efficiently, but more importantly because under the communist dictatorships citizens could not organize to protest against environmental degradation and were unable to move the state machinery to take effective action. There is little doubt that the Western capitalistic democracies have a far better record in this context. In this system, capitalist enterprises, often in league with the state, as under the Reagan administration in the United States of America (1980–8), do try to externalize the cost of environmental degradation as much as possible. However, where this personally hurts large numbers of people, as with air and water pollution, citizens can force the system to take effective remedial action (see Hays 1987). Equally importantly, market forces do tend to promote efficient resource use by private enterprises. Thus while Indian and Russian industries have remained among the least efficient in the use of energy, Scandinavian and Japanese industries are by far the most efficient.

THE CAPITALIST DREAM

Clean and efficient technologies have been favoured in some capitalist countries, thereby promoting more effective conversion of natural to human-made capital. The need to reduce adverse environmental impacts has also induced substitution of information for material and energy; thus carefully crafted electronic devices perform a variety of functions far more effectively than heavier, unwieldy mechanical devices which consume more materials and energy for their fabrication.

While the capitalist system thus promotes more efficient use of resources it also continually encourages higher levels of consumption of material, energy and informational resources. Private enterprise thrives on ever-greater consumption of the goods and services it produces. In all capitalist societies people are sucked into a rat race, consuming more and more of the resources of the earth. However, this has by no means led to more satisfactory lives. As Ivan Illich points out, the availability of private automobiles in the United States has not reduced, but rather increased, the time an average citizen spends in daily commuting to work. The frenzy of earning and spending in the capitalist system is thus not at all compatible with moderating the impact of people on the environment (Durning 1992; Illich 1978).

But then many in the First World question the need for moderating the impact of people on the resources of the earth (Simon 1981; Simon and Kahn 1984). Global omnivores too see few if any signals of environmental degradation hurting them at a personal level. This is for the same reason that Indian omnivores have little concern with environmental degradation, namely, that they have successfully passed on these costs to the ecosystem people and ecological refugees of the Third World. Europeans and North Americans drive automobiles fabricated in Japan made of iron mined in India. The devastation of the Indian hinterland stripped of ore from open pits is not a signal that reaches them in any form. Nor is the pesticide pollution and resultant extinction of several species of river fish endemic to the eastern Himalaya at all evident to European consumers sipping cups of Darjeeling tea. The forest cover of Japan is among the most extensive in the world, around 60 per cent of the land surface, only because the Japanese can import and lavishly use and throw away wood coming from southeast Asia. The suffering inflicted on the ecosystem people of these forest tracts is emphatically not a signal reaching an average citizen of Japan. It is primarily through the operations of capitalism that the global sisterhood and brotherhood of omnivores can effectively shield itself from the larger environmental consequences of its actions.

With Marxism on the retreat and Gandhism never having acquired any real popular following, the philosophy of capitalism is today in the ascendant. In India too, economic liberalization is at the centre stage of the national debate on development. The proposals for liberalization call for a loosening of bureaucratic control and opening up of the economy to the outside world by the abolition of tariff barriers and restrictions on the operation of foreign capital. What implications does this programme have for the six root causes we have identified as lying behind the degradation of India's environment?

(a) The world capitalist system thrives on passing on the costs of environmental degradation to the ecosystem people of the Third World; the Indian version of economic liberalization is therefore quite unlikely to enhance the access of India's ecosystem people to natural resources.

(b) Economic theory claims that, under ideal conditions, market forces result in an efficient pattern of resource use, which leads in turn to the maximization of the level of satisfaction of the parties concerned. So in theory ecosystem people and ecological refugees too should, under a system of privatization, enjoy greater access to the human-made capital of India. The reality is far from such an ideal. In India's grossly inequitable society the market assigns a far lower weighting to demands by weaker segments of the society, so that resources flow towards production of commodities in demand by the omnivores. Moreover, the market is so manipulated that labour or commodities which the weaker segments have to offer are obtained from them at a very low value, with organized trade and industry usurping the larger share of profits accruing from economic transactions.

(c) The philosophy of economic liberalization advocates pruning the bureaucratic apparatus, doing away with state subsidies, permitting market-driven competition to operate more freely. These are all measures that ought to enhance the efficiency of conversion of natural capital into human-made capital. But piecemeal application is unlikely to achieve this to any significant degree. Today, the size and powers of the Indian bureaucracy are hardly being pruned. Subsidies are being only selectively pared down; thus, cutting down on subsidies for the supply of water to industry, to city dwellers or to rich farmers is a subject conspicuously absent from policy debates. Nor are private enterprises being compelled to act in an environmentally responsible fashion. This does not augur well for long-term improvements in the efficiency of resource use in India.

(d) Economic liberalization is not likely to cut down the power of omnivores to capture the country's capital of natural resources; rather, this power may increase as indigenous omnivores ally more strongly with those of the First World.

(e) Economic liberalization will not put a brake on the omnivore's appetite for, and access to, human-made capital either, especially as omnivores are likely to hold on to their ability to pass on the costs of production to the ecosystem people and ecological refugees of the country.

(f) Economic liberalization may further accelerate the drain of the country's natural resources abroad, unless it sufficiently strengthens capabilities of exporting products of manufacture or high-technology services.

The philosophy of economic liberalization does have at its core three significant themes: namely, the need to do away with state-sponsored subsidies; to prune the powers and size of the state apparatus; and to create an open, democratic society that should in theory imply a better deal for India's environment and people. Implemented in the context of a highly inequitable society, however, this philosophy is unlikely to lead to any genuine progress.

A WORKING SYNTHESIS

Each of the three contending philosophies – Gandhism, Marxism and liberal capitalism – thus has components which are very desirable when viewed from an environmental perspective. But each philosophy also has components that deserve to be decisively rejected. What is evidently needed is a synthesis of the several positive elements: decentralization and empowerment of village communities along with a moderation of appetite for resource consumption from Gandhism; equity and empowerment of the weaker sections from Marxism; and an encouragement of private enterprise coupled to public accountability in an open, democratic system from liberal capitalism. In so far as Gandhism seeks to conserve all that is best in our traditions, it might be called the Indian variant of conservatism, with this significant caveat: that it seeks to conserve not the hierarchy of aristocratic privilege, but the repository of wisdom and meaning vested in ecosystem people. Marxism is of course the best-known strand within the socialist movement, while democratic capitalism is the ideology of the liberal. If our arguments are correct, then the environmental philosophy most appropriate for our times is in fact, nothing but conservative-liberal-socialism.

The alternative development paradigm flowing out of conservative-liberal-socialism would have the following elements with respect to the key issues identified by us (see Table 2, p. 111):

1 India should move towards a genuinely participatory democracy where the political leadership as well as the bureaucracy is made accountable to the masses of people. This requires the strengthening of grassroots democracy, conferring substantial powers on *mandal panchayats* and *zilla parishats*. It would also be desirable to reconstitute the existing states of the Indian union into smaller, more homogeneous units, breaking up huge provinces like Uttar Pradesh and Bihar. Following Gandhi's vision the country might come to be made up of self-governing village republics with no powers available to higher-level political authorities arbitrarily to dissolve or suspend elections to their *panchayats*. The *zilla parishats* and state assemblies should be similarly protected against arbitrary interference from higher levels. This nurturing of widely participatory democratic institutions should be complemented by an opportunity for people to decide directly on a wide range of developmental issues. In this context the system operating in the US state of California might be appropriately adapted to Indian conditions. Under this system any issue may be put on the ballot at the time of election provided that a minimum number of constituents so demand it, by attaching their signature to a petition. Thus the voters in California have decided against installing any more nuclear power stations through a public referendum at the time of state elections, a decision then binding on the state government. Analogously, an appropriate constituency of Indian districts might be empowered to decide on whether they want high dams in the Himalaya after a debate has exposed them to the merits and demerits of various alternatives.

2 The process of control, planning, implementation and monitoring of natural resource use should be radically restructured to render it an open, democratic process with full public accountability, and with substantial powers of controlling the resources of each locality being devolved to the local population. This calls for a pruning of bureaucratic authority and a transfer of most of its powers to local grassroots-level democratic institutions. Thus forests, grazing lands and irrigation tanks should revert to management by local communities, on their own or (as in the case of joint forest management, which is discussed in Chapter 7) in partnership with the state. The government should be relieved of its draconian powers to acquire land and water resources without the consent of the local people, without paying due compensation, or without appropriate arrangements for resettlement.

Instead, the local people should be empowered to work out plans for developing natural resources and managing local environmental affairs in a manner fine-tuned to the specific local situation and in accordance with their aspirations. Such a system would permit a much more fruitful utilization of the traditional knowledge and wisdom possessed by ecosystem people. The programmes of natural resource development so formulated should also be implemented largely by local people, closely supervised by them to ensure proper public accountability. In all this process there should be full freedom of access to and sharing of information. The local educational institutions should become publicly accessible repositories of environmental information pertinent to their own localities and should play a key role in planning natural resource use working with local communities. Higher-level political authorities should not drain away all the profits arising from use of local natural resources; a substantial fraction of such profits must be ploughed back to motivate local people to husband local resources prudently .

This is not to advocate that natural resource management should become an affair exclusively grounded in small-scale village-level programmes totally controlled by small communities. Larger-scale enterprises would undoubtedly continue and require proper co-ordination at larger scales. What is suggested, however, is that such programmes should be designed not by riding roughshod over local communities, as happens today, but through their involvement and after winning their consent. This would again require full freedom of information to the public, as well as public appraisals of the social, economic and environmental consequences of all development projects, whether they be on the scale of a small irrigation tank, or a series of large dams on the River Narmada. The public should also be involved in developing a detailed environmental management plan for each locality, ensuring that no undue damage is inflicted, even on a smaller scale, in rushing through large projects. Monitoring of project performance should be made mandatory and involve the public, again to ensure efficient execution sensitive to the local environment and the welfare of the local people.

3 Decentralization and wider public involvement can also put a stop to the widespread undervaluing of natural resources. In numerous localities of the southern states of Karnataka and Andhra Pradesh, for instance, the quarrying of granite has emerged as a lucrative business. Leases are granted by the state government to private operators, without taking into account the wishes of the villagers in whose vicinity these lands lie. But granite quarrying has contributed greatly to environmental degradation, through deforestation and the blockage of streams used for both drinking water and irrigation. At present, the concessionaires of stone and sand quarries pay a very small royalty, and that too to the state. So they have little interest in careful, efficient use of the resources. Rather they would blast or dig away as rapidly as possible, quickly carting away whatever they can. If the resource was

properly priced, and a goodly share of the profit were to come to the local communities, they would ensure much more prudent use – as indeed happened in the case of the Baliraja dam mentioned in Chapter 2. There the villagers claimed the rights over the royalty from quarrying sand in the river bed. They then ensured that the sand was quarried in such a way as to help in digging the foundation of the dam they wanted built, rather than haphazardly disrupting the river course as had been the practice of the contractors. Thus an important reform that environmental movements and grassroots-level politicians must press for is the proper valuation of natural resources, with a substantial share accruing to local communities from the profits realized from the utilization of these resources, whether timber, granite, coal or oil.

A proper price tag must also be attached to degradation of the environment, charged directly to the agent responsible. Thus villages suffering crop loss due to pollution from a cement factory must have the power to levy an appropriately scaled tax on the industry to compensate it for the loss. Only then would the 'polluter pays' principle be properly implemented and lead to effective action. In addition, the government should collect an environment management cess on a range of economic activities that have environmental consequences. This cess should not go into the kitty of the central or state governments, but directly, without any attenuation, to the village *panchayats* covering the length and breadth of the country. The *panchayats* could then use these funds to organize appropriate monitoring of environmental degradation, through a combination of local ecological knowledge, basic information collected through schools and colleges, or by other technical non-government organizations and research institutions hired at the discretion of the *panchayat*. This could create an entirely new support base for generating detailed, locality-specific scientific information that would be of great value in then deciding on the pricing of natural resources within *panchayat* territory, or on the level of the environment tax to be imposed on offending industries. This information (and revenue) could additionally be put to use in designing a development strategy that would at once benefit local communities and be gentle on the environment.

4 We are quite clear that such a development process could succeed only in a far more equitable society than India is presently. Halting the pace of environmental degradation, then, depends on progress towards a more equitable access to resources, to information and to decision-making power for ecosystem people and for ecological refugees. In India's predominantly agricultural society, a key reform to promote equity is radical land reform. At the same time we need to reverse the current inequitable pattern of flow of state-sponsored subsidies – a flow which is at present biased towards the better-off omnivores, at the cost of the weaker sections. Industry, chemicalized agriculture and urban islands of prosperity must all pay a fair price for the resources they utilize and must bear the full cost of treating the pollutants

they discharge into the environment. The utterly inefficient bureaucratic apparatus that presides over this function of mobilizing resource fluxes in favour of the omnivores should be dismantled. This in itself would cut down significantly on the ranks of omnivores. The state should then pass on the finances thus saved to the decentralized political institutions at village and district levels, for the local inhabitants to use in accordance with their own priorities. Evidence suggests that both health care and education would be a high priority were people given a free choice (in the state of Kerala, where popular participation in the political process is more intense and rewarding than in other parts of India, universal literacy and excellent health care facilities have indeed been its consequence (Jeffrey 1992)). However, the content of education should be broadened to emphasize first-hand observations of people's own surroundings. Investments in health care and in meaningful education linked to the development process, and creation of decentralized political institutions along with land reform would move Indian society towards a far more equitable condition. Since the interests of ecosystem people as well as ecological refugees are more compatible with the prudent use of natural resources, such moves in the direction of economic and political equity would promote a people-oriented as well as environmentally sensitive process of development.

5 The collapse of communist regimes in eastern Europe and the recent shifts towards private enterprise in China are striking proof of the failures of state socialism. Here the Indian experience, with its corrupt, inefficient and wasteful bureaucratic apparatus, is entirely in consonance. It is clear therefore that we should continue the shift towards encouraging private enterprise on all fronts, for producing goods and services which people would pay for on the market, while increasingly providing social services such as health and education through the voluntary sector. While freeing the private sector from undue restrictions such as licensing, we must also withdraw all state-sponsored subsidies, adopting instead pricing policies to promote environment-friendly behaviour. State subsidies have so far merely encouraged exhaustive, wasteful resource use and focused the attention of entrepreneurs on manipulation and bribery, rather than on delivering goods and services in an efficient fashion. Pollution regulation should be strongly enforced, but should not remain a business carried out in secret by official agencies, since such an arrangement promotes corruption. A culture of efficient resource use and control of pollution by industrial enterprises would not only serve the cause of environment, but would also enhance the competitive ability of these enterprises in the international marketplace.

Private enterprise should also be encouraged in delivering services such as education, health care or watershed-based soil conservation for which communities might pay through public funds. India's experience of delivering such services through the state apparatus has been very disappointing indeed.

Once guaranteed a salary, many state-employed teachers or doctors simply do not go and work in remote villages. By contrast, voluntary agencies have often done an excellent job of delivering education and health care in rural areas. This is primarily because the voluntary-sector workers are not guaranteed permanent jobs regardless of how good their performance is. They have to compete with each other and to demonstrate continually to funding agencies that they are indeed fulfilling their tasks. At the present time, voluntary agencies play a minor role limited mostly by the availability of foreign funding. Instead, internally generated funds must devolve to local communities, which should have the option to pay for such services only if they are delivered efficiently.

6 The appropriate scale of economic enterprises, especially of development projects, has been a major subject of controversy. Proponents of 'small is beautiful', Gandhians and appropriate technologists have challenged the overwhelming bias in development projects towards the large scale. This preference has undoubtedly favoured the concentration of benefits in the hands of a small number of people. However, the opposition to big projects, for their own sake, is not always productive. It is not so much the scale of enterprise as the way it is conducted that is at the heart of the deprivation of ecosystem people and the degradation of their environment. Carefully treating individual watersheds, and building a series of small dams, may indeed be, in a technical sense, a more optimal solution than building one huge dam. However, watershed treatments may well be carried out in an even more inefficient fashion than the construction and management of larger dams, in which case little is gained by merely choosing projects on a smaller scale. On the other hand, if local communities can get together, experiment with alternative soil and water conservation techniques, cropping patterns and water requirements, and have an opportunity to consider a whole range of technological options before making an informed choice, a proper scale is far more likely to be selected.

It is very likely that today the choice of economic enterprises is unduly biased towards the large scale. But the solution lies not in merely correcting the choice in terms of scale, but in putting in place new mechanisms of deciding upon what projects should be chosen, while ensuring that they are implemented in an efficient and properly accountable fashion.

7 Technological advances thus far have supported a continual increase in the scale of economic enterprises: bigger dams, giant power projects and larger ships. Technological advances have also permitted larger resource fluxes from more remote areas. These developments have permitted a small proportion of the world's population to capture resources at the cost of the rest of the people. Technocrats have often been in league with the small number of omnivores, turning a blind eye both to social deprivation and to environmental degradation. Many environmentalists, especially those of a

Gandhian persuasion, have therefore come to reject technology altogether, dreaming of a return to an idyllic pastoral-agrarian system (Nandy 1989). But pre-colonial, agrarian society in India was beset with its own set of evils, including untouchability and the oppression of women. It is not at all clear if it was indeed in the idyllic state of Ram Rajya as some Gandhians believe it to have been. But that apart, control over technology is a potent force of domination in the world today. Any society that turns its back on advances in technology will quickly find itself exploited and subjugated by others.

Thus, the opting out of scientific and technological advance, which after all is a thrilling adventure of the human spirit, is not a route that India could possibly follow. What is needed instead is to look for ways in which this advance could be directed away from the current path of speeding up the drain of the country's natural resources, producing more polluting substances and concentrating power in the hands of an ever narrower elite. We believe that this can be achieved by taking advantage of the tremendous possibilities of rapid communication opened up by modern technologies to put in place a genuinely participatory democracy with decentralized political institutions endowed with decision-making power. Indeed, we are now very close to a situation in which the role of elected representatives can be severely pared down, with more decisions taken through broader-based referenda. Modern technology also favours greater access to information on the part of a wider public. For instance, with satellite pictures beamed to the earth and readily available for a modest sum of money, the forest cover or patterns of siltation of river beds in any part of the country can easily be assessed by an interested citizen. Such information can also be rapidly analysed and transmitted to thousands of other interested parties. Our emphasis should be on adopting a more open, people-oriented pattern of development that would counter the other, negative implications of technological advance.

This is a crucial theme that is elaborated in the next chapter. But we may note here that the country-wide 'people's science' movements, pioneered by the 25-year-old Kerala Sastra Sahitya Parishat, are attempting just this. They are involving people in assessing the implications of ongoing development processes, in what is happening to their environment and health, and in campaigning for positive alternatives. Demystifying science and technology, making it accessible to all people and involving them in deciding how it should be fruitfully employed is obviously the direction in which we should proceed; to turn our back on modern science and technology (as some radical Greens would have us do) is neither desirable, nor within the realm of practical possibility.

8 Environmental change is clearly related to human demands on the resources of the earth; these demands have been escalating both because of increase in human numbers and because of rapidly increasing per capita demands. The increase in numbers may be largely attributed to the ecosystem

people of the world; while the increase in per capita resource demands has largely occurred because of the omnivores, whose own numbers have been growing at a far slower rate. Now the transition to smaller families will take place only when parents can fruitfully invest substantially in enhancing the quality of their offspring, in equipping them to compete in the market for skilled labour. Ecosystem people would thus become motivated to rear a small number of offspring only when a more equitable development process gradually draws them into the ambit of modern industry, services and intensive agriculture. But with India's huge population the resource demands of such a large number of people in these modern sectors may simply become too large a drain on the country's resources. Industrial nations today get away with their own huge demands by passing on the consequences to Third World countries. This option can be open to only a small minority of the global community; it is certainly not open to India as a whole. The massive expansion of the resource demands of Indian omnivores has been enabled only through a type of internal colonialism. If India is to move towards a sustainable pattern of development, then the omnivores will have to accept restraints on their own consumption levels.

The Gandhian philosophy proposes that this limitation can be brought about through voluntary restraint. But this is impossible to put into practice. What one must urge instead is that omnivores should not be permitted to capture resources at the cost of ecosystem people and ecological refugees; they must bear such costs themselves. This calls for a discontinuation of state-sponsored subsidies benefiting the rich, a compulsion to take care of pollutants, and a democratic dispersion of powers to decide on the use of the country's resources. At the same time, however, the level of resource consumption of India's population need not be brought down or remain stagnant, for the present pattern of resource use is incredibly wasteful. Enhancing its efficiency can greatly enhance availability of resources and services that the resources provide for all of the country's citizens. There are also enormous possibilities here (as we show in the next chapter) of bringing this about through an information-intensive process of development: that is, the substitution of information for energy and materials that modern science is swiftly making possible.

9 During the colonial period India suffered grievously from a drain of its natural resources, while serving as a market for goods of British manufacture at adverse terms of trade. The Second World War greatly improved India's position *vis-à-vis* Britain, and on independence concerted attempts were made to protect India from the drain of its resources by creating tariff barriers against imported goods. This created a sellers' market, actively helped by state-imposed restrictions on production through licensing. On top of this Indian industry was pampered by state subsidies in the supply of natural resources. As a result, industrialists have concentrated on making huge profits through

obtaining licences and government subsidies by manipulation and bribery, while using imported, if often obsolete, technologies. They have completely neglected to use resources efficiently, or advance technologically. This lopsided industrial development has created a high-cost, low-quality economy unable to hold its own in the global marketplace. Meanwhile, India's demands for foreign exchange have gone on soaring with the increasing consumption of petroleum products and military hardware. That has led India into a debt trap, from which the only escape route is to beg for further loans by agreeing to demolish the barriers keeping foreign business out of India.

Where should India go from here? The solution favoured by Marxists, to renege on the international debt and to continue behind closed doors, is quite unworkable, unless the country is able drastically to cut down on its import bill. That would call for a new strategy of far more dispersed development of agriculture and industry requiring much lower levels of energy inputs, as well as a matrix of international relationships that would permit a deceleration of investment in the defence sector. Both of these developments would be highly desirable. But if they were to come about, then little would be gained by keeping out of global trade, provided only that India does take good care that this trade does not impose undue pressures on its resource base.

We have outlined above the broad parameters of a new development model for India, which we have called (only partly in jest) conservative-liberal-socialism. Subsequent chapters more closely investigate the implications of this model for scientific advance, for the management of forest resources and for the urgent task of stabilizing India's population. While we deal only with three sectors in this book, on account of both space and our own spheres of competence, our broader framework might be fruitfully applied to other key sectors of Indian development, such as energy, water utilization, transport and housing. But to conclude the present discussion, let us reaffirm that our alternative path of development would effectively address the six root causes of environmental degradation in India. Thus:

(a) It would confer on ecosystem people a substantial degree of control over the country's natural resource base. With the quality of their own life linked to the well being of the local environment, this is likely to lead to a far more prudent use of nature.

(b) It would create conditions under which ecosystem people would gradually come to share equitably in the country's capital of human-made resources. India cannot indefinitely maintain a dual society with the masses of people at subsistence level; only when these people enter the modernized sectors of industry–services–intensive agriculture will they become motivated to invest in the quality of children and produce a small number of offspring, thereby moving India towards a stable population.

(c) By drastically restructuring the management of natural resources, ensuring accountability and empowering segments of population with a stake in sustainable use, this development strategy would greatly enhance efficiency in the generation of human-made capital from natural capital.

(d) This strategy would restrict the level of access of omnivores to energy and natural resources. This is essential to put brakes to the current process of exhaustive use by omnivores, where the costs of degradation are passed on to ecosystem people and ecological refugees.

(e) This strategy would also temper the appetite of omnivores for consumption by ensuring that they have to bear the true costs of resource degradation.

(f) By reducing excessive dependence on external energy inputs, this strategy would put India in a far stronger position to organize trade with the outside world without putting excessive pressure on its natural environment.

An environmental strategy informed by conservative-liberal-socialism would be in the spirit of tolerance and assimilation of a diversity of strands of thought that is so characteristic of Indian culture. It would keep India alive as a country of vigorous, powerful communities controlling their own destinies to a significant degree. It would create a more open, democratic and egalitarian society which would also nurture socially responsible private enterprise. It would participate in the adventure of modern science and technology, putting it to use in the protection of the environment and in enhancing the quality of life of its people. It would permit India to integrate with the international community from a position of strength.

This is of course a strategy that would be opposed resolutely by the omnivores, themselves happy to liquidate the country's resource base, let it fall prey to a debt trap while continuing their own wasteful, inefficient ways. But it is a strategy very much in the interests of a vast majority of India's people. Moving in its direction would be a long and arduous, but by no means hopeless, struggle.

6

KNOWLEDGE OF THE PEOPLE, BY THE PEOPLE, FOR THE PEOPLE

THE DINOSAUR GOES UNDER

India has been and remains a biomass-based civilization. By this we mean that a majority of Indians depend on biomass gathered by their own labour, or produced through low-input agriculture to meet most of their subsistence needs. They also exchange such biomass, at best processed through simple manual labour, to acquire other materials and services they consume on the market. According to the 1991 census, fully 74.3 per cent of the Indian population is rural based, and most of these people depend on cultivating their own lands or working as labour on other people's lands for a significant fraction of their earnings. The Anthropological Survey of India (ASI) recognizes 4,635 communities as making up the Indian population. For each community they record as current occupations those in which a notable proportion of community members are engaged. According to the ASI, 5.1 per cent of communities are still involved in hunting-gathering, 2.4 per cent in trapping birds, 8.3 per cent in fishing, 3.7 per cent in shifting cultivation, 4.7 per cent in terrace cultivation in hills, 54 per cent in settled cultivation, 4.3 per cent in horticulture, 21.5 per cent in maintaining domestic livestock, 0.8 per cent in nomadic herding, 3.5 per cent in weaving mats or baskets, and 3.9 per cent in woodworking. Of these 4,635 communities at least some members of as many as 4,285 communities still collect fuelwood or crop residues for domestic cooking, while members of 2,514 communities collect dung for this purpose. In another study of eighty-two villages from the drier parts of India, it was found that 14–23 per cent of the earnings of all households was derived from collection of biomass from common lands around villages (Jodha 1990; K.S. Singh 1992; Gadgil 1993).

Of course, the subsistence of most of humanity was grounded in biomass before the beginning of the industrial revolution two centuries ago. But the industrial revolution brought to the fore the effective use of a range of new energy sources: coal, petroleum, hydroelectric power and most recently nuclear power and solar power. The availability of such substantial additional sources of energy has permitted the fabrication of an enormous body of material artefacts that now spread across the earth. To begin with, such artefacts were rather crude, guzzling energy and belching smoke like steam locomotives, or leaving ugly scars on the earth like open-pit mines. But with time they have become more efficient in resource use, and less polluting, as witness modern-day electric locomotives or mines that are carefully filled up and revegetated afterwards. Not that people are once again treading lightly on the earth. Far from it. But we are surely moving in the direction of economic processes that are much more efficient in energy and material terms, and we are doing this primarily by using more, and better, information. There is no doubt that humanity as a whole, having passed from a civilization grounded on biomass through one grounded on material artefacts, is now on course to erect a civilization grounded on information.

Thus far at least, Indian society has barely been touched by this transformation. Indeed, in a technological sense India is still a land of dinosaurs. Its heroic dams are among the largest of human-made objects, the web of high-tension power lines and irrigation canals fanning out from these dams among the most extensive of such networks. But the gigantic no longer commands the centre stage of world technology; rather, it is the diminutive that holds sway. It is as if the age of the ponderous dinosaurs were giving way to that of itsy-bitsy shrews. These tiny mammals may be seen, metaphorically at least, as having outstripped the behemoths on the strength of their brainpower. They could absorb, process and deploy information far better than could the cumbrous reptiles. Somewhat analogously, the minuscule artefacts taking over the world by storm are far more intricately constructed than the Titanics of yesteryear. These artefacts are, above all, devices capable of processing information very, very effectively. We are entering the age of hand-held video cameras, miniature computers, automated manufacture; of ways of absorbing, processing, deploying information to ever greater purpose.

The societies that dominate the world today – indeed, that have come to dominate the world over the past three centuries – have been those capable of handling information well. At the core of the modern scientific and technological revolution is a method of continually augmenting the social store of knowledge. This method involves an open sharing of information, continually relating it to the hard, verifiable facts of the world out there. No progress is possible unless the information is thus openly available, at least to the concerned social group.

It is thus no accident that democratic values have flourished hand in hand with the development of scientific knowledge. For science and technology can flourish only in an open society that encourages scepticism. The Soviet Union did for a while make some significant advances in fields such as space technology, but these were limited achievements. The same society fell far behind in other important disciplines such as genetics precisely because open discussion was suppressed, and because appeals to state authority or Marxist dogma, rather than empirical facts, came to dominate intellectual life. The closed Soviet society, in the tight grip of a political-bureaucratic elite, also fell back on two other crucial fronts: the electronic revolution that ushered in the modern information age, and adequate control of industrial (including radioactive) pollution. Focusing on heavy industry, Stalin ushered in the age of ponderous, polluting artefacts, an age of dinosaurs. This despotic society collapsed when the information age dawned in the more flexible, democratic societies of the West. It collapsed in good part because the Soviet military machine was handicapped by being so far behind in automatic control and guidance systems. Just as dinosaurs inevitably made room for the smarter mammals, so has European communism given way to a democratic, capitalist system.

Shaking off the colonial yoke with its suppression of information, India elected to follow the Soviet model of a centrally planned society, albeit in a democratic political order. Here it took from colonialism the state monopoly of information, and from the Soviet model a centralized, supposedly all-knowing bureaucratic-technocratic apparatus. This apparatus has since made all major decisions in secrecy, with no open discussions, no sharing of information with the people at large. As in the Soviet Union, this has very quickly degenerated into a system where a narrow alliance of omnivores has come to enjoy enormous power, including the power to suppress scrutiny to serve their own limited ends. At least under communist states universal education has been vigorously promoted, so that China has today over 62 per cent female and 84 per cent male literacy. Even this has not happened in India, so that it remains a country where half the population is illiterate and assetless, where all development decisions are made in a closed fashion, where disinformation to further the interests of those in power rules the roost.

This pattern of development has inevitably favoured the dinosaurs. When energy policy is being formulated, for example, the discussion always turns on the need to build more large power generation plants, be they nuclear, superthermal or hydroelectric, and on the amount of money to be sunk in these unwieldy projects with their long gestation periods. The discussion always bypasses a critical assessment of the tremendous wastage of power, of the large losses that state electricity boards have been incurring. There is no good basic information on simple parameters such as who uses how much of the electric power generated. Thus agricultural pump sets are not metered, and the pump set owners merely pay a small flat rate based on the horsepower of the pump. This makes it easy to arbitrarily assign some large power consumption to the pump sets to cover up for gross inefficiencies and irregularities of power use elsewhere. The electricity boards are widely suspected of routinely using this device to lower the figures of transmission and distribution losses. Even then these losses are officially of the order of 23 per cent; it is speculated that they are actually of the order of 30 per cent. This may be compared with losses in other Asian countries such as South Korea and Thailand, which are of the order of 12–15 per cent. In this manner, good information is simply not available, not just to the public, but with the electricity boards themselves. Bad management of information and indeed disinformation thus perpetuates a system in which enhancing the efficiency of energy use, which ought to receive a very high priority in a desperately poor country, is sidelined in favour of the building of more and more, larger and larger, less and less efficient power plants.

Small may not be inherently beautiful, nor large by definition ugly. But there are definite disadvantages associated with the large, given the long gestation between conception and actual commissioning. These long delays did not matter so much at a time when technology was changing at a snail's pace. It may have taken decades to construct the magnificent old Meenakshi temple in Madurai in south India, but building technology very likely stood still over the whole period. However, between the time we now start constructing a large hydroelectric project and its completion, twenty or thirty years later, solar power generation cells may well advance enough to make that a far more economic and environmentally desirable option. Keeping all options open and quickly taking advantage of new possibilities is therefore becoming more significant with every passing day in the modern world. A preoccupation with the execution of large, cumbersome projects is simply not in tune with the times. What is in tune is a spirit of remaining constantly alert, absorbing new information as it comes in, and putting it to good advantage in projects that can yield handsome returns before the technology they employ becomes thoroughly outdated.

PRACTICAL KNOWLEDGE

India has but one course open to it: to move forward towards an information-based society that generates for its large population the many services it needs, while sparingly using material and energy resources. This could permit us to retain intact, even replenish, the biomass that continues to be a basis of Indian civilization. Unfortunately the course on which we are currently launched at once abuses both biomass and information. That is why our forest managers massacre prime rain forests with claims of enhancing their productivity through raising eucalyptus plantations without any proper trials as to whether eucalyptus will succeed. That is why they find the eucalyptus plantations falling prey to the pink disease that subsequently wipes them out, converting into desert hill slopes once clothed by lush green vegetation (Sharma *et al.* 1984).

What would an alternative development strategy respectful of both biomass and information imply? It would involve putting into practice the approach of dealing gently with complex natural systems, of trying to enhance the services these systems provide for the people with many, small, timely inputs. It would focus on what people get, not on what they – or the state apparatus – spend. It would try to maximize the ratio of output of useful services to input of investment in terms of labour, money and so on. It would properly discount for any incidental loss of useful services as a consequence of the intervention. Since interventions in complex systems inevitably have unexpected, often undesired, consequences such a strategy would call for continual monitoring and feedback. Since our knowledge of the functioning of complex natural systems is limited, such a strategy would emphasize flexible, adaptive management. It would not call for rigid, inflexible, centralized plans, but locality-specific plans that are continually adjusted in the light of experience.

For India's biomass-based society, it is vital to ask the question as to who has the motivation, the knowledge and the competence to enhance the vital services the country's lands, forests and waters provide. Would it be ecosystem people, the local communities, who have served as stewards of these natural resources for centuries? Or would it be the bureaucracies, that have taken over as regulators over the last century and a half; or the manipulators, be they contractors involved in dam construction, or manufacturers of sugar or paper? And what of ecological refugees, such as peasants from the plains of Kerala encroaching on Western Ghats forests who fit into none of these categories of stewards, regulators or manipulators? It is, then, of interest to examine how these four categories rate with respect to motivation, knowledge and competence to manage the land and water resources of India, recognizing of course that none of the categories are homogeneous; that local communities are especially apt to be divided into many interest groups pulling in different directions.

Consider first the motivation for careful, sustainable use. Such motivation would depend on whether imprudent use today adversely affects the well-being in the future of the party indulging in misuse. One of the factors that could decide whether there is such a link is the size of the resource catchment: that is, the area over which the concerned parties garner resources. A social group with access to resources from an extensive area is unlikely to be motivated to use the resources of any particular locality in a sustainable fashion, for it would always perceive the option of moving on to another locality on exhaustion of the resources in any given locality. Thus manipulators like the commercial users of biomass resources of non-cultivated lands, be they forest-based industry, manufacturers of herbal medicine or suppliers of milk to metropolitan cities, deal with large resource catchments. They therefore tend to focus at any time on the resource elements that bring them the greatest profit, switching to other localities, or other kinds of resources, as these are depleted. Thus India's paper and plywood industries have always concentrated initial harvests in localities close to the mills; as these become depleted, they move on to localities further and further away. They also have options of switching to a different type of resource – for instance, to eucalyptus if bamboo is depleted. Manipulators with their vast resource catchments are therefore unlikely to be motivated to use the natural resources of any part of the country in a sustainable fashion (see also Gadgil and Guha 1992).

So far as the bureaucratic regulators and their political masters are concerned, their future wellbeing has no relation whatsoever to the prudent use of resources of any locality under their control. Given the present lack of accountability, with no rewards for honest performance as custodians, and no punishment for misappropriation of the resource base, the regulators stand only to gain from profligacy – except, occasionally, when a major misdemeanour comes to light and they are exposed to adverse publicity. One example is the case of the Coorg forests. Lying near the centre of the Karnataka Western Ghats, Coorg is an undulating plateau at an altitude of 1,000–1,500 m, famous for its coffee and cardamom estates and its martial traditions. The estates have extensive areas especially in the catchments of springs, traditionally maintained under natural forest cover. The planted areas also have large numbers of shade trees. Over the years the ownership of the land and the trees of Coorg has remained in flux with large tracts being neither under clearcut private ownership, nor under full control of the Forest Department as reserved forest. This has created a situation of uncertainty where constant shifts in government policy have ensured that the estate owners are tempted to dispose of the trees illegally and smuggle them across the state border into Kerala. A serious attempt was made to bring this difficult situation under control in 1991 at the personal behest of the Karnataka chief minister through a government order regulating tree felling. The order was apparently deliberately flouted for over six months by a group of government officials who issued licences for tree cutting in contravention of the order. This misconduct was brought to light in 1994 because a former Conservator of Forests of Coorg who temporarily had to leave the Forest Department for his protests against the working of the department was commissioned by the state legislature to investigate the whole matter. His report recommended that the Corps of Detectives be called in to investigate the malpractices indulged in by over fifty officials. It remains to be seen whether any of them are actually punished. But unfortunately few such cases are ever exposed. With little accountability, the regulators as a class have indeed no motivation to strive for responsible resource use, although there are undoubtedly a minority of public-spirited officials who do attempt to enforce prudent use in the broader national interest.

What of the local communities, then, divided along many lines of class and caste? With increasing penetration of the market, the sizes of resource catchments of local communities are rapidly expanding; more and more of what they use is being brought to them from further and further away. Thus, throughout India, a large number of poorer rural people depend on weaving baskets and mats to supplement their earnings outside the agricultural season. But as bamboo, reeds and other materials have been locally exhausted, such activities increasingly depend on import of raw material from considerable distance and marketing of the produce in distant towns and cities. However, the near-total dependence of large numbers of poorer rural people on locally collected plant material for fuel and fodder continues to this day. Since women are especially involved in gathering of such biomass, they are particularly concerned with maintenance of sources of such material locally, while men, more strongly involved with the market economy, are less concerned. This, for instance, has been the experience of the Dasholi Gram Swarajya Sangh in the Garhwal Himalaya. This organization, which pioneered the Chipko movement, primarily operates through conducting week-long eco-development camps in village after village. During such camps people of surrounding villages come to help work on building protective stone walls and to undertake plantation activities. The local villagers collectively cook and feed the entire volunteer corps. Over the years women have taken an increasingly active role in these camps, as they did in the early protests against deforestation. As recounted in Chapter 4, a recent study of satellite pictures showed that these camps have played an important role not only in halting the pace of deforestation, but in helping rebuild the tree cover of the Alakananda valley.

But over much of rural India, and increasingly even in the tribal areas of the northeast, there are members of local communities whether from old large landholding families or the newly prosperous political bosses who depend little on the natural resources of their own localities. Rather, they are happy enough to see these liquidated if it means a fast buck for themselves. These contradictions are reflected in some of the experiences of social forestry programmes, for instance in Ranebennur *taluk* in the semiarid parts of Karnataka state. In several villages of this region the poorer segments of the local communities wanted the produce of these plantations distributed among residents at concessional rates, while the village council president, coming from a richer family, wanted these auctioned on the open market.

So, while a significant fraction of India's rural and tribal communities does favour long-term, prudent use of local natural resources, there are elements among them with no such motivation. However, the actual behaviour of local communities pretty much uniformly throughout the country tends to short-sighted, exhaustive use. This is because these communities today have almost no control over their local resources. Except in parts of northeastern India, such control was taken over by the state, through its Forest, Revenue or Irrigation Departments. With this state takeover local communities have no rights to control use of the common property resources, either by their own members or outsiders, although they may enjoy certain privileges of limited use. Consider, for example, the recent failure of the villagers of Halakar in the Uttara Kannada district of Karnataka to stake a claim on the timber harvested from their village forest. As mentioned in Chapter 2, Halakar was one of the three villages specially earmarked for praise by Collins, the British official who was appointed to look into the settlement of forest lands of the district in 1921. As a result of Collins's recommendations a village forest committee for Halakar was formally established in 1928, and has continued to function effectively to this date. It has overcome the arbitrary dissolution of the committee by the state Forest Department in 1968, winning a court case against this order. So there is no doubt that the villagers of Halakar deserve full credit for the current good state of forest cover in their village forest. Now, part of this land was acquired by the government for the West Coast railway line in 1992, with the Forest Department stepping in to organize clear-felling of this area. The villagers then claimed rights over the timber, but failed to sustain their claims; the timber was taken over by the Forest Department.

All over the country such usurpation of rights over resources by the state at the cost of local communities ensures that people fear that fruits of their own prudence will go to somebody else, and therefore have little motivation to exercise restraint. The situation is even more precarious with the ecological refugees, with no roots in the localities they migrate into, and with a very insecure control over the resource base. Under the current regime, then, no segment of Indian society is motivated to use the common-property natural resources of the country in a prudent fashion, least of all the state apparatus that has monopolistic control over it. The only exception is in parts of northeastern India, where local communities do retain a measure of resource control and occasionally do exhibit prudent behaviour, as for instance in the case of the establishment of safety forests by the Gangtes of the Churchandapur district of Manipur discussed in Chapter 4 above.

Prudent management of natural resources also calls for knowledge and capability along with motivation. The technocracy, be it foresters, agricultural scientists or civil engineers, claims monopoly over this knowledge. But these are dubious claims, because these resource managers are dealing with highly complex natural systems. The behaviour of a complex natural system depends on many subtle interactions among the diverse components making up the system. Thus bamboos are among the most prominent and useful constituents of India's tropical forests. They have been used by rural people for centuries, as house construction material, for weaving baskets, for fabricating fishing or agricultural implements. In recent years bamboo has come to serve as an important raw material for the paper and polyfibre industry as well. As noted in Chapter 2, the state has made bamboo resources available to the industry at throwaway prices, and bamboo resources have declined rapidly following industrial exploitation. The question is the role of many different factors impinging on bamboo stocks in their decline. One of us was involved in an investigation of this problem between 1975 and 1980. We worked in the district of Uttara Kannada, in forest areas leased out to the West Coast Paper Mill. The paper mill's forest department, headed by a very experienced retired forest official, prescribed the clearing of the covering of thorny branches that forms at the base of any bamboo clump. This was in the belief that this facilitated good growth of the bamboo culms. It turned out that the practice exposed the new bamboo shoots to grazing by langur monkeys, wild pigs and cattle. Our field investigations over three years conclusively demonstrated that this was a major factor responsible for the failure of new culms to develop, resulting in the depletion of bamboo stocks. The local people had also been harvesting bamboos for their own use for centuries. They never cleared this thorny protective covering of young shoots. Indeed, we later discovered that they were aware of the problems caused by removal of the thorns. But the forest department of the mill had never consulted them – nor for that matter did we until well after the investigations were launched. In consequence the mill was paying its labourers to carry out operations that actually hurt the resources they valued.

Ecosystem people have for centuries depended on the natural resources from a limited resource catchment to provide them with manifold services. They have therefore discovered a spectrum of uses for the local natural resources, be they soils or rocks, plants or animals. Omnivores, on the other hand, have been able to draw resources from vast areas, and to process them to provide many different services. They have therefore tended to force any given locality into exporting one particular resource, in the process writing off a whole range of other possibilities. Thus the British attempted to convert much of the forest of peninsular India into single-species stands of teak (*Tectona grandis*), a tree that was valued first for shipbuilding, and later for furniture and house construction. In this single-minded pursuit of teak plantations they ring barked and killed a wealth of tropical rain forest trees. A century later, the forest managers of independent India decided that the tropical forests of peninsular India must now produce eucalyptus as pulpwood, and in pursuit of this goal clear-felled vast stretches of rain forests. Notably, in both cases monocultural plantations often failed because of lack of attention to locality-specific factors (Gadgil and Guha 1992).

Ecosystem people know of a vast number of uses of the many plants of tropical rain forests; not just trees, but also herbs, shrubs, climbers, the epiphytes. One such useful plant is a herb, *Rauwolfia serpentina*, which produces an alkaloid reserpine useful in the treatment of high blood pressure. This herb, endemic to the Western Ghats, has been a part of the tribal medical repertoire for a long time; this information was picked up by practitioners of modern medicine some twenty years ago. Soon a commercial drug extracted from *Rauwolfia* came on the market, triggering off rapid exploitation and near extinction of the herb before its export was banned.

Ecosystem people have been in the business of extracting services from nature without large inputs (primarily because they had no access to them) for a very long time. Their practices have therefore been moulded to working closely with nature. This repertoire includes a great variety of land races of cultivated plants and domesticated animals adapted to particular environments which often are reservoirs of valuable genes conferring resistance to diseases, permitting salt or drought tolerance and so on. These land races are currently being rapidly lost, not necessarily because more productive varieties have been introduced, but often simply because of government pressures for acceptance of a package of modern practices carrying with it inducements of subsidies for supply of fertilizers, pesticides and so on. Such subsidized packages have tended to destroy many desirable practices, for example the desilting of irrigation tanks and the use of silt as a manure, or the use of plants for insect control, for example leaves of *Strychnos nuxvomica* in betelnut orchards of Uttara Kannada district.

A SCIENCE FOR SUSTENANCE

The practical knowledge and wisdom of India's ecosystem people must therefore once again come to assume an important role in enhancing the services being provided by natural systems, a role that is today being completely denied. We are, however, by no means proposing that modern science and technology should therefore withdraw. Folk knowledge is particularly relevant for processes that are evident to the eye and that take place on time scales of less than a few years. It does not extend to microbes that may be polluting drinking water sources, or to processes of soil erosion that are manifest only over decades. Enhancing ecosystem services beyond traditional levels must therefore depend on wise use of newer understanding and technologies. Indeed, in coming decades the startling new capabilities of moving genetic material at will from one organism into another are bound to revolutionize the whole process of biological production for human use. It would be folly not to take full advantage of such developments.

As important as these technical developments is the application of the scientific methodology in the endeavour to enhance ecosystem services. At the heart of the scientific methodology is the careful recording of empirical observations, the development of a model of how any natural system functions on the basis of such observations, the predictions of how it may behave in response to specific interventions, the verification through further observations of whether indeed changes take place as predicted, the incorporation of appropriate changes in the model of the system behaviour, and so on. This is an open-ended process through which the understanding of natural processes and human abilities to intervene effectively progresses step by step. Of course, from time to time such a process yields a spectacular new understanding, and often such understanding permits entirely new and highly effective forms of human intervention. Such was the appreciation that matter and energy are interconvertible, an understanding that permitted the release of atomic energy. Such also is the appreciation of the chemical nature of genetic material, which is now permitting the creation of genetically engineered life forms. These landmark developments are of course important; but every one of these major advances in understanding, as well as the development of techniques based on them, has taken place outside India. Importing these techniques could bring considerable benefits to India; but that should not be the sole context in which India employs modern science and technology. Rather India must get down to applying the scientific methodology across the length and breadth of the country in enhancing ecosystem services, not necessarily in spectacular leaps and bounds but in slow, steady increments.

This the current system utterly fails to accomplish. The irrigation engineers take care neither of the catchment nor of the distributories, so that dams silt up at far greater rates than projected, while large parts of lower command areas fail to receive any irrigation water. Soil conservation measures are prescribed without considering the nature of the substrate, and are so sloppily executed that they often enhance, not reduce, rates of soil erosion. The agricultural strategy promoted by the agricultural universities and state departments prescribes standard, rather heavy, doses of fertilizers and pesticides with no reference to the field-to-field variation in many relevant factors such as levels of soil nutrients, or factors that vary from day to day such as levels of pest populations.

While natural resources are everywhere treated in this sloppy fashion, the technocratic agencies make bogus claims of scientific management. These have no substance, for these agencies do not maintain any careful data on the state of the system, they do not monitor whether their interventions have indeed had the projected consequences, and they do not adjust their operations to correct for any deviations between the projected and realized outcomes. Rather, state agencies employ the jargon and prestige of science to cover up for the stark inefficiency and wastefulness of the technocratic management of India's natural resources (see Gadgil and Guha 1992; Singh 1994).

An alternative strategy of providing good information inputs evidently needs to be put in place. This should be a system attuned to expose rather than obfuscate what is happening to the resource base; that would look for ways to minimize external intervention, rather than hunt for excuses to spend state funds; that would try to work out a programme of timely, small, appropriate interventions to make the most of the potentialities of variable natural systems. Such a programme of injecting scientific inputs would not be glamorous and its results would have very limited international recognition. Nor would the prescriptions generated create large opportunities of profit for omnivore enterprises. In consequence the sophisticated, largely American-trained, leadership of the Indian scientific community has little interest in such an endeavour. Neither, of course, has the state technocracy, which has unfortunately developed vested interests in wasteful resource use. It is instead the 'people's science' movement groups that have so far promoted such exercises, albeit in small ways. Perhaps the most notable example of such an exercise comes from the Baliraja dam movement of the peasants of the village of Tandulwadi in the drought-prone Sangli district of Maharashtra, to which we have already alluded in Chapter 2. To recapitulate, the peasants of this village decided that the standard package of using irrigation water to grow a limited repertoire of water-hungry crops on a small area by a few farmers was inappropriate. They then worked closely with a voluntary group called Mukti Sangarsh and a highly qualified irrigation engineer, K.R. Datye, to work out a strategy that would not simply maximize commercial profits per hectare of land under unlimited irrigation, but rather would maximize the total quantity of additional biomass produced as a result of irrigation. Such a strategy favours deployment of irrigation water in more limited quantities in any given area, but its more uniform use over a much wider region. To work out such a strategy calls for careful experiments with different crop mixes and different schedules of irrigation. The strategy has to be flexible to permit effective use of small quantities of water in years of drought, with a shift to more water-demanding crops to take advantage of abundant rains in good years. Furthermore, the crop mix is to be so chosen as to produce varied biomass components to meet the manifold requirements; onions to be marketed as a cash crop, sorghum for consumption as food, subabul to provide green manure, eucalyptus to generate wood for local landless artisans, and so on. Obviously this is a challenging scientific problem, albeit of specific application to a particular locality. It is rather like the Japanese way of doing things – a process of deliberate, patient, continual improvement in productivity; every year making the product a little better, the process a little more resource efficient. It is time Indians applied this Japanese philosophy to the way the natural resources of the country are managed and tried slowly, patiently, continually to enhance their productivity. That is surely the key to success in the long run.

This would call for a radical restructuring of the scientific effort pertaining to the management of the country's natural resources. Today this is a narrow-based, bureaucratic effort pursued through centralized institutions. Even where there has been an attempt to set up dispersed research centres, as with the agricultural universities, there is little genuine understanding of conditions on farmers' land. True, some trials are carried out in farmers' fields, but the farmers themselves are scarcely involved in deciding on the contents of the experiment. The agricultural scientists by and large have little respect for people's knowledge; they are also bent upon pushing a package of uniform practices involving heavy, standardized applications of water and agrochemicals. At the extreme is forestry science, dominated totally by bureaucrats who take an occasional year or two to take on research assignments. The research is conducted in a thoroughly non-academic atmosphere with results published in house journals where acceptance is related to bureaucratic status and not scientific merit. Nor is the research usually related to what is happening in real forests 'out there'. Thus a hundred years after the Forest Research Institute was founded, the National Commission on Agriculture had to remark that it was quite unable to decide on whether the area controlled by the Forest Department was 69 or 75 million ha (National Commission on

Agriculture 1976). And after continually pointing for decades to rural biomass needs as the main cause of the many failures of forest management, the Forest Department has not conducted a single careful empirical study of this problem.

A bottom-up research strategy involving the wider masses of people would look to them to pose research problems of relevance. This would follow if the local communities are permitted to reassert control over the resource base. Then the communities would begin to ask for scientific inputs that would benefit them in a sustainable fashion, creating a genuine demand for environment-friendly science and technology. This demand could be effectively met by a much more decentralized network of institutions of learning. The undergraduate colleges, some 6,000 of them scattered throughout the country, could come to play a critical role in this endeavour. Today these colleges are merely involved in teaching routine, unimaginative courses with little relevance to real life. But they have a large body of students and teachers who could be stimulated to take up the challenge of investigating the status of the environment, to create a publicly accessible database on local natural resources, to continually monitor how various developments are impinging on it and to help work out ways of carefully, incrementally enhancing resource productivity through timely, appropriate inputs. A consortium of local community institutions, of high schools, colleges and research bodies would have to be organized to provide the necessary scientific inputs for this enterprise. That would be the organization of a science for sustenance, indeed a science for continuous progress.

Good management of natural resources would also call for the capability of continually monitoring and regulating such use. This is an enormous task given the vastness of the country and the complexity of its natural and social systems. It is no doubt true that modern technological advances, such as in remote sensing and computer-based databases, permit efficient collection and handling of large masses of information. But in spite of these advances, centralized planning of locality-specific interventions and monitoring of their consequences is an impossible task, especially when one is dealing with complex natural systems that vary greatly from location to location. The consequences of adding a certain amount of irrigation water to a field depend on the type of the soil, its slope, its aspect, the crops under cultivation, the weeds, insect pests, fungal diseases affecting the crops, the amount of manure or fertilizer used, the amount of money, labour available with the farming family, the human diseases prevalent, and so on. Equally variable would be the consequences of releasing carp seed in a freshwater pond, dependent in turn on the seasonal fluctuations in water level, the temperature regime, weeds, other fish present, eutrophication and chemical pollution, as well as on who owns the pond, who have been traditionally fishing in it and so on. Such detailed information would only be available in each locality at any point in time. Moreover, the values of relevant parameters can go on changing with time. For instance, a new insect pest may arrive; people may learn to like carp although they had looked down on them before; or there may be an exceptionally heavy downpour that breaks down the bund. Tackling all the changes in myriad localities in a centralized fashion is not a practical proposition, despite rapid improvements in capabilities of handling information. Instead, a far more cost-effective and efficient process of planning and management would be to carry out the component operations in a parallel fashion in each of the many different localities, restricting centralized functioning only to necessary co-ordination.

There are two very different ways of dealing with complex natural systems that are highly variable in space and time. The first is to try to fine-tune the inputs to the specific context, to gently nudge the system through many small, but timely and appropriate, inputs towards enhancing the services people want. The other approach is to ride roughshod over the complexity and variation, to homogenize the system through heavy external inputs and then manage it, employing some standard interventions uniformly prescribed for large areas. Consider, for instance, the problem of drinking water becoming ever more scarce in many Indian villages. One approach is to depend on local precipitation, however scanty, by properly regulating its flow, ensuring that it gets collected in some appropriate means of storage, protecting the stored water against pollution or bacterial contamination, and then using it in appropriate quantities at appropriate seasons. The other is to permit local water storages to get silted up or polluted, and then supply water to the village by bringing it over large distances in tankers, a solution that is indeed in vogue over large portions of the state of Maharashtra. The first approach calls for very careful local-level planning that can have some central scientific and technical inputs, but ultimately must depend on local motivation, knowledge and community-based management. The second approach needs little local planning and management, but can be run as a centrally planned, controlled operation. It uses a lot of external inputs – diesel for the tankers, for one. But that only creates a strong vested interest in the omnivore community in favour of the second, and far less desirable, approach.

Evidently, then, on all criteria – motivation, knowledge, management capabilities – the local communities are the most appropriate agency to look after and organize fine-tuned, prudent use of India's natural resource base, whether of forests, grazing grounds, irrigation tanks or agricultural lands. The state and the technically more sophisticated people of course have an important role to play, but that role should be of facilitation, co-ordination, of adding information not available to the local communities. Instead the state and its technocracy have assumed a very different role: that of agents of an extractive economy with no interest in prudent, sustainable resource use, controlling the resource base, claiming monopoly over all pertinent information and making all decisions on how to deal with the natural resources of the country.

How might we then move in the direction of strengthening the motivation, the knowledge base, the capabilities of the local communities towards prudent management of the country's natural resources? How might we transform the present predatory role of omnivores into one of mutualism with the ecosystem people of the country, paying a fair price for access to the natural resources, and contributing to their sustainable utilization through their stock of rapidly growing scientific knowledge and entrepreneurial abilities? The next chapter outlines the steps that might facilitate this transition in the crucial sector of forests and biodiversity conservation.

7

WHAT ARE FORESTS FOR?

THE FORESTRY DEBATE

In the global history of natural resource management, there are few institutions as significant as the Indian Forest Department. Set up in 1864, it now controls over one-fifth of the country's land area, backed by an impressive administrative and legal infrastructure. Not only is the Forest Department India's biggest landlord, it has the power to affect the lives of virtually every inhabitant of the countryside. In what is still dominantly a biomass-based economy, all segments of Indian society – peasants, tribals, pastoralists, slum dwellers and industry – have a heavy dependence on the produce of the forests, as the source of fuel, fodder, construction timber or industrial raw material.

And yet, in the century and a quarter of its existence, the Forest Department has been a most widely reviled arm of the Indian state. As narrated in Chapter 3, popular opposition to the workings of the Forest Department has been both sustained and widespread. One might say that underlying these varied protests have been two central ideas: that state control over woodland (as opposed to local community control) is illegitimate; and that the Forest Department's programmes of commercial timber harvesting have seriously undermined local subsistence economies. Since the early 1970s, grassroots organizations active in forest areas have called repeatedly for a complete overhaul of forest management. State policies, they contend, have excluded the dominant majority of the Indian population from the benefits of forest working while favouring the interests of a select group of industries and urban consumers. At the same time, they have asked for a reorientation of forest policy, towards more directly serving rural subsistence interests (see Fernandes and Kulkarni 1983). Nor are tribal and peasant groups the Forest Department's only critics. Thus conservationists have argued that commercial forestry has contributed significantly to the decimation of biological diversity and to an increase in soil erosion and floods. More recently, industrialists, who have hitherto been the prime beneficiaries of forest policy, have hit out at the department's partial withdrawal of subsidies and at being at last denied preferential supply of raw material. Faced with sharp criticism from several quarters, senior forest officials have claimed that they are unjustly being made scapegoats while the real causes of forest destruction escape identification (Misra 1984).

Despite nearly two decades of vigorous debate, however, policy changes in the forestry sector have been slow. Where earlier policies had narrowly focused on commercial exploitation for industry and the market, there is now greater talk of 'people's participation' and 'ecological security'. All the same, there is a lack of conceptual clarity – both within the government and outside – of how best to manage the country's forests in a sustainable fashion, while at the same time minimizing conflicts among the varied and often competing demands on its produce.

This chapter outlines the elements of what we believe to be an appropriate forest policy for India, one consistent with the alternative development agenda outlined in Chapter 5. Forestry is a resource sector to which we have devoted much of our own work. It is thus that we use forestry, rather than water or energy, to illustrate how the principles of conservative-liberal-socialism might be put into practice in one vital area of India's development efforts. Specifically, we show how the cardinal weaknesses of forest management to date lie in the failure to identify clearly the competing demands on India's forests. We then go on to prescribe ways of meeting these demands in a manner consistent with the imperatives of ecology, equity and efficiency – with clearly specified roles for the state, the market and the local communities respectively.

Let us first present the arguments of the key actors in the ongoing debate on Indian forest management. There are four important groups in this debate, whom we might characterize as wildlife conservationists, timber harvesters (i.e. industrialists), rural social activists and scientific foresters respectively. These groups are indeed 'interest groups' in the classical sense of the term: in other words, each has a specific claim on the resource under contention, and lobbies actively to defend and promote this interest. What is noteworthy is that in each case, the management proposals advanced by the group seek wider support from a sophisticated theory of resource use in which their own specific interests are presented as being congruent with the general interests of society as a whole.

Wilderness conservationists

We begin with wilderness conservationists, an interest group small in number but with a major influence on policy. While their practical emphasis concerns the preservation of unspoilt nature, defenders of the wilderness are prone to advance moral, scientific and philosophical arguments to advance their cause. Although the initial and possibly still the dominant impulse is the aesthetic

Plate 21 Tropical rain forests still cover extensive areas of the Western Ghats hill chain, identified as one of the world's eighteen biodiversity hot spots

Plate 22 The pristine rain forest of the Andaman and Nicobar Islands is a treasure house of biodiversity

value of wilderness and wild species, this sentiment has found strong support from recent biological and philosophical debates. The theme of biological diversity as an essential component of direct and indirect, known and yet to be discovered, survival value for humanity and an emphasis upon the intrinsic 'rights' of non-human species have been prominent in recent debates on wilderness preservation. The quite specific interests of nature lovers in the preservation of wilderness are thus submerged in the philosophy of 'bio-centricism', which validates strong action on behalf of the rights of non-human nature (Guha 1989b).

These philosophical claims notwithstanding, in India a select group of ex-hunters and naturalists has been in the forefront of wilderness conservation. Their concern has been overwhelmingly with the protection of endangered species of large mammals such as the tiger, rhinoceros and elephant. Their influence is manifest in the massive network of parks and sanctuaries, many of which are oriented around the protection of a single species, notably the two dozen or so parks under the celebrated Project Tiger. They share with senior bureaucrats in particular a similar educational and cultural background, and this proximity has in no small way influenced the designation and management of wildland. It is thus quite fair to characterize them, as we have done in Chapter 4, as being environmentalists with an 'omnivore' background.

Timber harvesters

The second important group in the Indian forestry debate consists of those who view the forest as a source of industrial raw material. In terms of their management preferences, timber harvesters are the polar opposite of the wilderness purists. While the latter stand for a 'hands off' management style that involves the minimum possible interference with natural processes, industrial demands on the forest often involve substantial and even irreversible modifications of natural ecosystems.

The industrial view of nature is simply instrumental. The forest is a source of raw material for processing factories, and the pursuit of profit dictates a pragmatic and flexible attitude towards its management. In the past, timber harvesters had been content with letting the state manage forests, so long as they were assured abundant raw material at rock-bottom prices; now, with increasing deforestation and the withdrawal of subsidies, wood-based industry has been lobbying hard for the release of degraded forest lands as captive plantations. Yet, in response to the environmentalist challenge of the past two decades, some industrialists have been quick to develop their own general theory of resource use. On the one hand, to justify their claims on public land they continually invoke the equation in conventional development thinking of industrialization with 'progress' and prosperity. At the same time, they argue that captive plantations will significantly lessen the

Plate 23 Having economically exhausted the timber resources of mainland India, forest-based industry is now increasingly dependent on the as yet little-exploited forests of the Andaman and Nicobar Islands

pressure on natural forests, whose destruction, largely at their own hands, they had hitherto been indifferent to.

Rural social activists

By 'rural social activists' we refer to those individuals and groups who work among ecosystem people dependent on the forest for a variety of economic needs, both subsistence and commercial. In India this would include groups of hunter-gatherers, shifting cultivators, pastoralists, artisans, landless labourers, and small, medium-sized and big farmers. A large proportion of the rural population lives close to a biological subsistence margin, and access to fuel, fodder, small timber and non-timber forest produce is critical to their ways of making a livelihood. Moreover, most Indian villages have stayed on one site for centuries, and the collective consciousness of their inhabitants stretches far back into the past; state usurpation of the forest is from this perspective a comparatively recent phenomenon, and thus resolutely opposed by tribals and peasants, who cling tenaciously to traditional conceptions of ownership and use.

These deep-rooted animosities are invoked by rural social activists in their polemic against state forest management. Some among them call for a radical reorientation of forest policy, towards more directly serving the interests of subsistence peasants, tribals, nomads and artisans; others go further in asking for a total state withdrawal from forest areas, which can then revert to the control of village communities which, they claim, have the wherewithal to manage these areas sustainably and without friction. These specific recommendations draw sustenance from a powerful philosophy of agrarian localism, namely Gandhism. Where Gandhi, the 'Father of the Nation', always gave theoretical and practical primacy to rural interests, the policies of independent India are indicted – often rightly – as being heavily biased in favour of urban and industrial interests.

Scientific foresters

We come finally to the group in actual territorial control of forests and wildlands. The brief of the Forest Department is the adjudication of the competing claims of the three interest groups dealt with above, and this in a 'scientific' and 'objective' manner. Historically, scientific foresters saw themselves as heralding the transition from *laissez-faire* to state-directed capitalism, in which they were, along with other professional groups, the leading edge of economic development. Conservation is for them the 'gospel of efficiency', and scientific expertise and state control its prerequisites. These ideological justifications aside, in practice foresters have often tended to act on the basis of narrow self-interest, tenaciously clinging to control over all forest areas and the discretionary power that goes with this control (see Lal 1989).

In this manner, through the skilful submergence of specific interests in a general theory of the human and natural good, these competing groups have legitimized their actual claims on forests and woodland. The territorial aspirations of foresters are advanced by claims to a monopoly over scientific expertise; the aesthetic longings of nature lovers are legitimized by talk of biological diversity and environmental ethics; the profit motive of capital masquerades as a philosophy of progress and development; and the demands of agrarian populations are juxtaposed to an ideology of the rights of the 'little man'.

Finally, let us note the varying positions on state control over forests and wildland. Two groups are unambiguous in supporting state control over the commons, even if they insist that the state enforce only their definitions of forest use. Wilderness conservationists see an interventionist and powerful state as indispensable in both designating wildlands and keeping out intruders, while for scientific foresters state control is virtually a *sine qua non*, in that it allows scientific experts the room to plan rationally and at a nationwide level. Agricultural interests, for their part, while clearly expecting the state to take their side against other interest groups, are by and large opposed to state forestry, arguing instead for community ownership and management.

Plate 24 Hills in dry parts of peninsular India have been laid totally bare by charcoal production to meet urban demands followed by overgrazing and hacking for fuelwood by villagers

Lastly, industrialists are characteristically opportunistic about the question of forest ownership, calling for privatization of forest land when it suits them, and for state control and subsidized raw material when it does not.

These four contending groups apart, a fifth category of resource users has also exercised a major influence on the direction of Indian forestry. This consists of the urban middle and upper classes, who constitute a substantial and growing market for a variety of forest produce. Unlike the other interest groups in the forestry debate, urban consumers, while numbering in the tens of millions, do not have a co-ordinated perspective on forest policy. None the less, their demands for paper, plywood, quality furniture and processed non-wood forest produce have powerfully stimulated processes of forest destruction in many parts of the country. Ironically, the upper class also constitutes the constituency from which wilderness conservationists spring and to which their arguments are largely addressed. Where on the one hand the consumption ethic of the urban elite contributes to forest exploitation, on the other hand their aesthetic and recreational preferences help determine the priorities of park management. Squeezed between these twin processes of destruction and conservation are the vast bulk of the rural population, who have little stake either in commercial forestry or in wilderness areas as presently managed.

HARMONIZING CONFLICTS

The functions that forests perform can be classified under five major heads:

(a) maintenance of soil and water regimes;
(b) conservation of biological and genetic diversity;
(c) production of biomass for subsistence (i.e. as fuel, fodder, agricultural implements, building materials, etc.);
(d) production of woody biomass for commercial purposes (i.e. as raw material for industry); and
(e) production of non-timber biomass (e.g. cane, sal seeds, tendu leaves) for commerce.

The fulfilment of functions (a) and (b) may be said to constitute the *ecological* objective of forest management; of (c) the *subsistence* objective; and of (d) the *developmental* function. As for (e), in so far as it generates substantial employment among local (especially tribal) communities it constitutes both a subsistence and developmental objective.

In terms of our interest group framework, one can see immediately that wilderness conservationists have been concerned overwhelmingly with function (b) (and, to a much lesser extent, with (a)); rural social activists primarily with (c) and to a limited extent with (e); and timber harvesters almost exclusively with (d). These varied, often conflicting, demands are then made on the state agency physically in command of forest land. Indeed, since its inception the Indian Forest Department has had primary, and often sole, responsibility for assuring the ecological, subsistence as well as developmental objectives of forest management. Moreover, under its management system of working plans, these competing demands are often placed on the same territorial area or patch of forest.

In practice, however, the developmental objective has come to assume overwhelming importance in forest management. As we have documented in some detail elsewhere (Gadgil and Guha 1992: Chapter 6), the industrial thrust of state forestry has severely undermined subsistence options by restricting access of villagers to forest areas, and contributed greatly to deforestation and allied ecological degradation (see also Centre for Science and Environment 1985; Fernandes and Kulkarni 1983). Clearly, an alternative forest policy must separate out these three objectives – in a conceptual, territorial and management sense. Following from this premise, the following discussion offers distinct policy options and institutional frameworks for fulfilling the ecological, subsistence and developmental objectives of forest management.

THE ECOLOGICAL IMPERATIVE

India has a great diversity of environmental regimes. In the northeast it can boast of areas with the highest levels of annual precipitation in the world; at the same time parts of the Thar Desert may see no rain for years on end with the annual average being below 300 mm. Its northern plains may experience temperatures of 50°C in the summers, while the higher reaches of the Himalayas remain perpetually snow-bound. India contains some of the highest mountains of the world as well as tiny coral islands in the Indian Ocean. Its river systems include the mighty Ganga and Brahmaputra with their huge flood plains and the short and swift west-flowing rivers discharging from the Western Ghats into the Arabian Sea. In consequence, India's natural vegetation ranges widely over tropical evergreen and mangroves to dry deciduous and desert scrub.

Indeed, India qualifies as one of the top twelve countries in the world in terms of biodiversity because of its tropical and subtropical climate, its great variety of environmental regimes, and its position at the junction of the African, Palaearctic and Oriental regions. It is also a part of a secondary, diffuse centre of domestication of plants and animals. Conserving the Indian heritage of biodiversity is a major challenge, especially given the large human population with its subsistence needs, and the growing resource demands of the urban–industrial sector.

This effort must consider biodiversity along many dimensions and on many scales of organization, ranging over the genetic, species, ecosystem, landscape and regional levels. At any of these levels, we must consider the

Plate 25 Earthen mound inside which an unusual bird, the Nicobar megapode, lays its eggs. India, one of the world's top twelve countries for megadiversity, harbours a rich variety of plant, animal and microbial species

Plate 26 This fig tree, possibly new to science, is one of the many species of trees endemic to the Andaman and Nicobar Islands

relative value of the different elements; after all, we set a higher conservation value on elephants and tigers than on the smallpox virus. In general, the rarer elements would be valued higher, as would elements of greater economic promise such as the wild relatives of cultivated plants. Other aesthetic, scientific and moral considerations would also enter into setting conservation priorities. In particular, if a major justification for conservation is the 'transformative value' of wilderness (Norton 1987) – that is, its value in moulding human attitudes towards the environment as a habitat for humanity – then we would decide in favour of the biodiversity being widely dispersed rather than concentrated in a few large reserves.

Thus far, efforts at protecting diversity have focused almost exclusively on large nature reserves. These cover about 4 per cent of the country's surface; not all of this constitutes 'core areas' under strict protection. That is, it also includes areas from which timber harvests may be continuing (see Kothari *et al.* 1989). These large reserves have tended to be managed in the interests of the more spectacular larger mammals such as deer and tiger, although now there are reserves aimed at conserving other groups such as orchids. The philosophy underlying the management of nature reserves has focused on the banning of hunting by all segments of the population, and exclusion of subsistence demands by the local rural and tribal populations. Thus long-settled villages have been shifted out of nature reserves, sometimes leading to damaging conflicts (see Chapter 3). Cultivators living on the periphery of these reserves are inadequately protected against crop damage and man-slaughter by wild animals, and very rarely compensated (see Sukumar 1989). The formal state-sponsored conservation effort has also been indifferent or inimical to traditions such as the belief in sacred groves and the protection of monkey populations.

Little thought or effort has gone into the conservation of biodiversity outside the nature reserve system. Indeed, the management regime prevalent outside the category of reserve forests has been an open-access regime in which no segment of society has a long-term interest in sustainable resource use. Development policies in general have also tended to promote exhaustive use of resources, both in reserved forests and outside them. Two elements that seem to be particularly significant in this context, as we showed in Chapters 1 and 2, are high levels of subsidies to the resource users and the lack of accountability on the part of resource managers. The pros and cons of doing away with resource subsidies, injecting responsibility into the actions of resource managers, including possibilities of participation in management by groups outside the state bureaucracies, and finally transforming the open-access regimes currently responsible for the tragedy of the commons are key issues that must be debated widely.

The current emphasis on nature reserves as the tool of biodiversity conservation, to the exclusion of all else, flows out of the overall development strategy. As we have shown, this strategy has equated development with the intensification of subsidized flows of resources to urban–industrial–irrigated agriculture sectors. These sectors have come therefore to acquire an increasingly larger share of economic and political power to the exclusion of the masses of peasants dependent on rain-fed agriculture and the small tribal population dependent on forest resources. It is these latter ecosystem people who relate directly to natural environments and depend on a diversity of biological resources for their personal wellbeing. On the other hand, those in power are well shielded from any direct negative consequences of all forms of environmental degradation including the loss of biodiversity. At the same time, they successfully monopolize most of the benefits flowing out of developmental activities. For them, therefore, environment and development often appear in contradiction, with environmental interests being naturally the ones to be sacrificed. Biodiversity conservation tends to be low on the priorities of the development strategy moulded by them, usually being restricted to nature reserves with a recreational appeal for the elite.

This development strategy explicitly attributes all environmental problems to the population pressure of the large masses of the ecosystem people directly dependent on the natural resources of their immediate environment. Therefore the exclusion of these people from nature reserves by the official machinery of the state has become a major focus of the state-sponsored conservation effort.

An examination of the adequacy of this approach is an appropriate starting-point for our discussion. This may be looked at from two perspectives. From the perspective of biological processes, the focus on nature reserves means that eventually all land or water areas outside this system are likely to be reduced either to highly degraded desert-like vegetation or to monocultures of just a few species such as eucalyptus, *Acacia auriculiformis* or a small number of high-yielding varieties of crops such as rice, wheat or tobacco. There may additionally be a few bodies of water dedicated to the production of carp or shrimp. The net result would be the persistence of a few species-rich fragments of a restricted range of habitats in the matrix of a biologically poor landscape and waterscape. Ecological theory tells us that such fragments are bound to lose a large proportion of their species in the long run. Such loss is all the more likely in the event of the global warming that may materialize over the next few decades.

It may be important, in this context, to strive to maintain, or better still re-create, a biologically richer overall matrix in which the species-rich fragments of nature reserves will be embedded. This would enhance the possibilities of long-term persistence of biodiversity in two ways. First, this matrix, which may include up to 40 per cent of the country's land surface under natural vegetation in various stages of degradation through human intervention, could on its own support a fair diversity of species of either commercial or subsistence value; this could easily run into several hundred species of trees, climbers, shrubs, herbs and grasses. These economically valued species would in turn support a large number of other species, especially invertebrates and smaller vertebrates. In addition, a small proportion of land in such a matrix, perhaps 5 per cent, could be devoted to biodiversity conservation in the form of species-rich islands such as individual fig trees or groves ranging from a few hundred square metres to a few hectares in size, to constitute a system of highly dispersed patchy refugia. This could end up adding another 2 per cent of the country's land surface to the nature reserve system. Such a matrix would also be of value in facilitating the dispersal from one reserve to another of many species at present restricted to the larger nature reserves. Ecological theory tells us that long-term persistence of species in any given locality is related to the rate of immigration into that locality of individuals of the different species present in the overall species pool. Such immigration would become all the more critical if global warming sharply increases rates of local extinctions. All in all, the maintenance of a biologically rich matrix with a system of small refuges serving as stepping-stones for dispersers would be a highly desirable complement to the system of nature reserves.

The cultivated lands and human habitations could also play a useful role in the maintenance of biodiversity through offering very small refuges such as the traditional banyan and peepul trees that dot India's villages and towns. Farm bunds, stream and river sides, canal sides and roadside avenues could all contribute towards the maintenance of a biologically friendly matrix of conservation areas.

The second important issue to be explored is how society could organize the maintenance of biodiversity in the system of nature reserves, as well as in the matrix of non-cultivated vegetation in which the nature reserves would be embedded. The current approach is one in which technical decisions as well as policing are in the hands of the state apparatus. One may, in this context, ask three crucial questions:

(a) Does this apparatus have at its disposal the information needed to make appropriate decisions?
(b) Is this apparatus adequately motivated to maintain biodiversity?
(c) Is this apparatus competent to carry out effectively the task of regulating human interventions in the interest of the maintenance of biodiversity?

There is room for questioning the adequacy of the state apparatus on all three counts. The personnel of this apparatus primarily derives from the Forest Service. The bureaucracy in general, including the forest bureaucracy, has monopolized the collection and interpretation of information on natural resource management in India. However, this has been done in a most casual and unscientific fashion, so that the scientific competence of this apparatus is in grave doubt. Moreover, ecological systems are exceedingly complex systems. The behaviour of such systems can be predicted only to a very limited extent on the basis of general principles. At the same time, historical observations of the system behaviour are a valuable input into predicting the outcomes of human interventions in specific systems. This implies that knowledge of folk ecology available with the local people may compare to and even exceed in value the knowledge base available with the official apparatus (see Chapter 6).

Second, no member of the official apparatus has a personal stake in the maintenance of biodiversity in any given locality. Of course, the local people may or may not have such a stake depending on how different elements of biodiversity affect their personal wellbeing. But if they receive direct benefits from maintenance of biodiversity for themselves (for instance, through supply of wild fruit or herbal medicine or protection of stream flow) or if they are rewarded by the wider society for protecting biodiversity, they are far more likely to be motivated to maintain the biodiversity of the localities to which they are intimately attached.

Finally, the state apparatus is quite ineffective in discharging a policing function unless it has full local co-operation, as has been strikingly brought out by the experience of the failure to apprehend the notorious sandalwood and ivory smuggler of Karnataka, Veerappan. There is, therefore, every reason to believe that the system of biodiversity conservation could be made much more effective by utilizing local folk ecological knowledge, and by creating a stake for the local population in conservation of biodiversity, if necessary by a system of payment of monetary reward as service charges (Gadgil and Rao 1994).

A decentralized system of biodiversity conservation, which might come to cover all of rural India, provides a great opportunity for developing a symbiotic relationship between systems of folk knowledge, traditional knowledge systems such as Ayurveda, Siddha or Unani medicine and modern scientific knowledge. A great deal of locality-specific knowledge of biodiversity elements significant to their own lifestyles resides with ecosystem people, many of them illiterate. This, for instance, is the case with specialist fisherfolk, who have an intimate knowledge of water bodies and ongoing changes in them and their snail, bivalve, shrimp, crab, fish fauna; or nomadic shepherds, who know a great deal about large tracts of scrub savannas and grasslands and their vegetation. But this knowledge can and should be tapped to feed into a wider process of biodiversity conservation. Some pioneering attempts at documenting such knowledge have already been initiated through an organization called SRISHTI under the leadership of Anil Kumar Gupta at the Indian Institute of Management, Ahmedabad. SRISHTI has organized a series of biodiversity contests with the help of primary schools in different states, which have discovered everywhere exceptional individuals, children as well as adults, who know hundreds of local species. SRISHTI also runs a network called HONEYBEE for giving due credit for and sharing such knowledge of distribution of biodiversity, as also its uses, often newly discovered by various innovators.

As we visualize it, rewards for effective custodianship of biodiversity and knowledge of its use would flow primarily to a geographically defined community; though they may also go to individuals, to caste or tribal groups or to clusters of village communities. It is important for the rewards to be in the form of assertion of community rights over public lands and waters within their defined territory. With the communities standing to gain in the long run they are likely to organize sustainable use patterns for these lands and waters, and to manage them in such a way as to also enhance their biodiversity value. However, it is absolutely essential that they should have adequate authority to exclude outsiders, and to regulate the harvests by group members, as well as an assurance of long-term returns from restrained use for such a system to operate effectively.

Such additional rights of access to publicly held resources would serve as a positive incentive for making prudent use of public lands and waters to meet local biomass needs. But the decentralization of natural resource management may by itself be inadequate to promote maintenance of high levels of distinctive elements of biodiversity within the community, since there may be better economic returns from monocultures, be they of high-yielding crop varieties on private lands or eucalyptus on public lands. Specific incentives, which should be viewed as service charges, are therefore necessary to maintain diversity, whether of cultivars on farm lands, of indigenous livestock breeds, of fruit trees in homesteads, of medicinal plants, wild relatives of crop plants, or troops of primates or crocodiles on public lands and waters. Individuals or communities participating in such efforts must therefore be paid certain rewards linked to the levels and value of biodiversity within their territory. Such rewards could be untied funds coming to the community to be devoted either to community works such as educational or health facilities, or to be shared among all community members. The rewards could also take the form of building community capacity for maintaining enhanced value of biodiversity within their territory, or for setting up biodiversity-based enterprises, such as chemical prospecting or extraction of active ingredients. Similar rewards may also flow for making available knowledge pertinent to uses of biodiversity, for instance in pest control.

Apart from these rewards, which might provide sustained positive incentives to custodians of biodiversity, there could be one-time rewards such as fees for collecting some genetic resource from the territory, or fees for sharing some piece of knowledge relating to use of biodiversity. There may also be shorter-term rewards such as royalties from commercial application of some element of biodiversity or some piece of knowledge relating to its use. It will, however, be very difficult to channel royalties of this nature properly to particular individuals or communities, since this would require every such element of biodiversity or knowledge to be traceable to a particular set of localities, communities or persons. It might therefore be better to pool such royalties in a national biodiversity fund and use this for rewarding communities for the ongoing maintenance of biodiversity within their territories (see, for details, Gadgil and Rao 1994). In this manner, community efforts at biodiversity conservation would innovatively supplement and perhaps in the long term largely replace more conventional schemes for the protection of wild areas and wild species through national parks and sanctuaries.

THE IMPERATIVE OF EFFICIENCY

The industrial orientation of state forestry in India has passed through three stages (see Gadgil and Guha 1992). In the first phase, *c.* 1870–1960, traditional selection felling methods of timber harvesting were relied upon in the belief that industrial demands could thus be met in perpetuity. However, for a variety of reasons, including a poor understanding of the ecology of species-rich tropical forests, an inadequate database, no proper planning for local subsistence demands and corruption, the objectives of 'sustained yield' forestry have been honoured mostly in the breach (see also Food and Agricultural Organization 1984). In the second phase, *c.* 1960–80, selection

Plate 27 Raft after raft of timber float down many streams of the Andaman and Nicobar Islands

felling was supplemented by large-scale clear-cutting of natural forests, which were then replaced by monocultural plantations, chiefly of exotic species such as eucalyptus and Caribbean pine. In the third phase, *c.* 1980 to date, wood-based industry has turned increasingly to purchasing raw materials from tree farmers or through imports.

As numerous studies have shown, commercial tree felling and the raising of species-poor plantations have proved to be most incompatible with the ecological functions of natural biological communities. At the same time, by largely excluding village communities from its fruits, industrial forestry under state auspices has also seriously violated the imperative of social equity. Not surprisingly, wood-based industry and the forest departments that tend to serve its interests have been the prime target of environmentalists.

As the proposals we are about to offer are likely to be controversial among those very environmentalists, a few clarifying remarks are in order. In their often justifiable attacks on the insatiable appetite and ecological insensitivity

Plate 28 India's forest-based industry is now tapping the rain forests of the Andaman and Nicobar Islands

of the forest industries sector, environmentalists, particularly those of a Gandhian persuasion, have sometimes tended to argue that India should turn its back on industrialization altogether. Undeniably, luxury furniture and consumer goods (such as chocolate and shampoo) based on the processing of select non-timber forest produce are hardly central to the developmental goals of the country. But the same cannot be said of paper, rayon, packaging materials, tendu leaves, etc., materials indispensable to the welfare of hundreds of millions of Indians.

If we are to reject, as we must, the extreme option of closing down all industrial units, the question remains – how best can we assure the raw material requirements of forest-based industry consistent with the imperatives of ecology, equity and efficiency? There are four alternative paths to choose from:

(a) continue to vest this responsibility with the Forest Department and its methods of timber harvesting from public land, with the caveat that subsidies be withdrawn and industry be made to pay a fair ecological price for forest produce;
(b) rely increasingly on the import of wood and wood pulp, thereby shifting the burden of deforestation on to other countries;
(c) turn over state forest land directly to industrial units as 'captive' plantations; or

(d) consolidate and promote the efforts of individual tree farmers, and effectively link them to regional markets and industrial units.

Given India's ever precarious balance of payments, the second alternative is hardly desirable. There is little doubt that industrialists would themselves prefer alternative (c), and have thus for several years been lobbying vigorously with state and central governments to hand over land on long lease (see Jain 1989). As for the Forest Department, notwithstanding the serious failures of 'sustained yield' forestry in the past, it would most likely prefer to continue these methods, under its control, of supplying raw materials to processing units.

Ecological considerations apart, from the viewpoint of efficiency too the state machinery is quite inappropriate for undertaking economic productive functions (indeed, studies have clearly shown tree farming by individuals to be far more productive, and far less costly, compared to government plantations). At the same time, conceding the claims of industry for captive plantations would inaugurate a new *zamindari* system, thereby violating the principle of equity. From all points of view, farm forestry appears the most feasible path for meeting industrial demands. As compared to state forestry, or industrial plantations, or indeed imports, it will distribute benefits far more widely among the population. Phasing out all commercial wood production from state lands would also work, all other things remaining equal, to promote the protection of natural forests.

Why then are so many environmentalists bitterly opposed to farm forestry? Morally, their opposition stems from a larger disdain for commercial transactions of any kind which go against the Gandhian ethic of local autonomy and self-sufficiency. Ecologically, they view eucalyptus – overwhelmingly the dominant species on the farm – as an unmitigated evil, for it is believed to poison the soil and lead to a lowering of the water table. Socially, they have indicted tree farming for increasing inequalities within the village, in particular by its displacement of food crops requiring more labour (see Bandyopadhyay and Shiva 1984).

Viewed dispassionately, the environmentalists' case against farm forestry, its strident rhetoric notwithstanding, rests on rather uncertain foundations. The Indian farmer needs cash income for a variety of purposes, from marrying off his daughters to making capital investments on his land. If tree crops are objected to in this regard, consistency requires that sugarcane, tobacco, cotton and a host of other commercial plant species (including cereals) be banished for the same reason. From an ecological perspective, the clear-felling of natural forests to replace them with eucalyptus (undertaken in many forest areas in the 1960s and 1970s) was clearly an unwise policy, but so far as farmland is concerned, eucalyptus is probably less harmful for the local environment than are other important cash crops such as sugarcane. The social criticism has more to commend it, though it too must be set against the gains in intersectoral equity (i.e. between town and country) represented by tree farming. Most substantively, farm forestry can only be assessed against the other options available to us – closing down all industrial units, harvesting wood from state forests, imports and the creation of captive plantations – among all of which it is incontrovertibly the most benign alternative.

Ironically, farm forestry has itself fallen on lean days. From the late 1970s, hundreds of thousands of cultivators, aided by the supply of free seedlings by the state, started planting eucalyptus and casuarina. According to one estimate, between 1981 and 1988 roughly 8,550 million trees were planted on farmland in India, with a survival rate of approximately 60 per cent. More than 80 per cent of these trees were eucalyptus. Yet by 1986 or so, disillusionment had set in, as yields and especially prices were well below expectations. In northwestern India and Gujarat – highly commercialized farming regions where eucalyptus appeared to have taken firm root – farmers were even uprooting saplings well ahead of maturity and replanting these areas with food crops.

Studies by N.C. Saxena of the Oxford Forestry Institute and by the Institute of Rural Management at Anand (Saxena 1990; Institute of Rural Management 1992) have highlighted several reasons behind this growing disenchantment with tree farming:

(a) lack of adequate technical information, in the absence of which farmers went in for high-density plantations with little spacing – this led to yields lower than anticipated, while the trees were too narrow for several commercial uses;

(b) depressed prices caused in part by competing sources of raw material for processing industry, especially subsidized supplies from forest land and imports;

(c) lack of market information on prices and sources of demand, leading farmers to depend heavily on middlemen, whose own margins were therefore much higher than the farmers';

(d) the existence in some states of restrictions on the harvest and transport of wood, compounded by the cost of obtaining permits from the bureaucracy; and

(e) in some cases, the lowering of yields of agricultural crops caused by declining soil fertility ascribed to the planting of eucalyptus.

These problems notwithstanding, long-time observers of farm forestry are agreed that it constitutes the most desirable option for generating raw material for industry. However, to make farm forestry remunerative and productive, a policy package with the following elements needs to be executed as soon as possible:

(a) The abolition of all laws and controls inhibiting individuals from planting or selling trees grown on private land.

(b) A moratorium on supplies of timber from state forest land and a phasing out of timber imports. A particularly poignant case is the recent eucalyptus glut in the northern state of Haryana, where prices fell during the 1980s from Rs 340 to Rs 190 per tonne. Ironically, in the neighbouring state of Himachal Pradesh excellent cedar forests are being felled for the manufacture of apple crates, which could just as easily be made from Haryana eucalyptus.

(c) These situations would be avoided when restrictions on timber harvesting from state land are coupled with efforts to link farmers directly with processing industry. Tree growers' co-operatives must be promoted to eliminate middlemen, and in support the state should stipulate that all wood-using enterprises (including those for perishable fruits like apples and oranges) be required to obtain their supplies from individual tree growers or their co-operatives.

(d) As a subsequent step, farmers' co-operatives might even set up processing units (e.g. small paper mills) to retain within the farm sector a greater proportion of value added. The milk and sugar co-operatives of western India might provide pointers in this regard (Attwood and Baviskar 1988).

(e) Scientific research needs to be undertaken on high-value, low-rotation tree species other than eucalyptus, that better complement food crop production.

(f) As is already the case for other cash crops, state bodies might generate and disseminate information on demand, supply, prices, etc., keeping farmers informed about prevailing market conditions.

(g) Finally, channels must exist for the flow of credit to tree growers, particularly to small farmers working on marginal land who might not otherwise take to long-gestation crops.

In conclusion, we should once again stress that shifting the burden of commercial wood production to private farmland makes eminent ecological sense. Owing to the greater productiveness and cost-effectiveness of farmers, the policies advocated here would require perhaps as little as one-quarter of the area required if we were to continue the traditional system of harvesting timber from state forests. By any reckoning, this constitutes an enormous potential saving of the country's natural endowment.

THE IMPERATIVE OF EQUITY

We come finally to the subsistence function of the forests, namely, the meeting of the varied biomass needs of India's villagers. Foremost among these needs are fuelwood and fodder, though access to small timber, thatch, green manure and raw materials for artisans such as basket weavers is also important for the livelihood of numerous villagers. Acute shortages of these materials prevail all over the country. While women have to walk longer and longer distances for fuelwood, and graziers to forage over a much wider area for fodder for their livestock, many wood-working artisans have been forced to abandon their calling altogether owing to the unavailability of raw material. Indeed, it is these shortages that lay behind the forest-based conflicts of the 1970s and 1980s (see Chapter 3). Fulfilling the biomass requirements of peasants, tribals, nomads and artisans in a sustainable and efficient manner thus constitutes perhaps the most urgent task of Indian forest policy. Ironically, the forestry report of the National Commission of Agriculture, a considered official statement on the subject, rejected these demands as illegitimate, arguing that villagers should depend instead on private lands and resources for generating biomass for their use (National Commission of Agriculture 1976).

Although some forest officials (and the odd wildlife conservationist) still maintain that state forest lands have no subsistence function whatsoever, this view is simply not valid. Demands of fuel and fodder, in particular, are too huge to be met in any significant proportion from private land. In 1981, for example, aggregate fuelwood demand in India was estimated at 262 million tonnes (mt), of which only 49 mt was being met from crop residues. While total fodder demand was estimated at 560 mt, the availability of agrowastes for this purpose was only 273 mt. Moreover, the bulk of crop residues are claimed by farmers with large holdings, who can in addition avail themselves of alternative fuel sources such as kerosene and cooking gas. But a majority of poor and landless peasants, nomads and artisans must perforce meet their biomass requirements from public land. The position with respect to thatch and building materials is undoubtedly very similar, whereas for a variety of non-timber forest products, villagers must turn *in toto* to lands owned not by individuals or communities but by the state.

On grounds of equity, therefore, the recommendations of the National Commission of Agriculture can be quickly cast aside, for the varied biomass needs of hundreds of millions of villagers can only be met, in the main, from lands presently constituted as 'reserved forests'. One could of course continue the present system, under which these demands are indeed fulfilled from state lands, but in a haphazard and unregulated fashion, and often by stealth. Another option, seriously advocated by some environmentalists (see Agarwal and Narain 1990) is to abolish state control totally, and let these lands revert to village communities to manage as best they can.

But where state control has lamentably failed to meet the subsistence imperative of forest policy, transferring reserved forests to the nearest hamlet would only, in the vast majority of cases, substitute one unregulated, inequitable and unsustainable system for another. This transfer might have the claims of justice – righting a historical wrong by returning forests confiscated by the state decades ago to those who previously owned them –

but it is only in exceptional cases, and under outstanding leadership, that village communities today can rise above divisions of caste and class, and the growth of individualism, to act consistently in their collective interest. (Their capacity to do so is also greatly undermined by the overall centralization of political authority that obtains over most of India.) All the same, we believe that a network of lands, constituted in the main from what are at present reserved forests, should indeed be totally devoted to meeting the diverse biomass needs of local people. Yet much more thought needs to be given to the appropriate management system, or systems, for these lands, in particular to the precise rights and responsibilities of the state and villagers respectively.

Valuable clues in this regard may be found in the experiences of two large experiments in subsistence forestry. One is the network of *van panchayats* (village forests) in the central Himalaya, the other the dynamic joint forest management programme of the West Bengal Forest Department. Although occurring in two widely separated areas with very different social structures, and originating in quite different historical processes, both systems offer us sharp pointers to the future direction forest management might take.

The *van panchayats* of the hill districts of Uttar Pradesh constitute the only major network of village forests, mandated by law, in India. They originated as a consequence of massive popular unrest at the state reservation of forests in the British-ruled Kumaun division. These forests were, for the colonial state, an important source of pine resin and timber; at the same time, they were situated in a region of great strategic significance, bordering both Nepal and Tibet and home to some of the bravest soldiers of the (British) Indian Army. Smouldering resentment at the state usurpation of forests culminated in a widespread peasant movement in 1921 that virtually paralysed the administration (see Guha 1989a). In its aftermath, the state decided to withdraw control over less valuable (from a commercial point of view) forest areas. Where in most other parts of the subcontinent the British were implacably opposed to promoting village-owned and -managed forests, in Kumaun they worked swiftly to allay discontent by allowing peasants effective control over large areas of woodland.

Under the Kumaun *panchayat* forest rules of 1931 (amended in 1976) a *van panchayat* can be formed, out of non-private land within the settlement boundaries of a village, on application to the deputy commissioner of one-third or more of its residents. Once formed, the *van panchayat* must elect its own managing committee of five to nine members, which then assumes control of the forest. Under the act, *van panchayats* have powers to regulate the use of the forest by villagers – e.g. the closure to grazing in certain seasons, restrictions on fuelwood extraction by individual households, and regulation of lopping – and levy fines on offenders. They can also prevent people from other villages from using their forest and prevent encroachments for dwellings or cultivation. Usually, a full-time watchman is appointed, paid from contributions by villagers. However, no timber trees may be felled without the permission of the Forest Department. The department is also empowered to give technical advice on timber and resin extraction. Of the receipts of *van panchayats*, 40 per cent are assigned to the Forest Department (for technical advice that is in fact very rarely given), 20 per cent is allotted to the *zilla parishat* (district council) while 40 per cent is kept with the deputy commissioner on account of the *van panchayat* itself, which with official permission may use this money for community services and local improvements.

As of 1985, there were 4,058 *van panchayats* in the Kumaun and Garhwal divisions, covering an area of 469,326 ha. Several studies have commented on the relatively healthy state of *panchayat* forests – usually in as good a condition as, and sometimes in even better shape than, reserved forests supposedly managed on 'scientific' lines – and this is corroborated by our own field experience. Oak forests in particular are invariably well maintained, though in some areas pine forests are also coveted by villagers for their grass. This is not to say that the functioning of *panchayat* forests might not be considerably improved. *Van panchayats* presently lack powers to collect fines directly from offenders (who may appeal to the deputy commissioner or go to court) or the ability fruitfully to spend money accumulated in their account on village improvements (thus in Ranikhet subdivision, Rs 3.8 million had accumulated unspent in the *van panchayat* account kept with the district magistrate). For its part, the government's *van panchayat* inspectors are too few and too poorly paid to fulfil their supervisory role properly, while the Forest Department does not feel obliged to provide technical advice as required under the act (Somanathan 1991; Ballabh and Singh 1988; Shah 1989).

These limitations notwithstanding, as an ecologically viable and socially equitable system of resource management, the network of *van panchayats* is a salutary reminder of the potential force of what often appears to be a tired cliché, namely 'popular participation'. A quite different system of popular participation in forest management has been more recently crafted in the state of West Bengal, far distant from the Kumaun Himalaya. In 1972, the West Bengal Forest Department recognized its failures in reviving degraded sal forests in the southwestern districts of the state. Traditional methods of surveillance and policing had led to a 'complete alienation of the people from the administration', resulting in frequent clashes between forest officials and villagers. Forest- and land-related conflicts in the region were also a major factor in fuelling the militant peasant movements led by the Naxalites. Accordingly, the department changed its strategy, making a beginning in the Arabari forest range of Midnapore district. Here, at the instance of a far-seeing forest officer, A.K. Bannerjee, villagers were involved in the protection of 1,272 hectares of badly degraded sal forest. In return for help in protection, villagers were given employment in both silvicultural and harvesting operations, 25 per cent of the final harvest, and allowed fuelwood and fodder collection on payment of a nominal fee. With the active and willing participation of the local community, the sal forests of Arabari underwent a remarkable recovery – by 1983, a previously worthless forest was valued at Rs 125 million.

Following the success of the Arabari scheme, village forest protection committees (FPCs) were started by the Forest Department in other areas. As of July 1990, there were 1,611 FPCs protecting 191,756 ha, primarily of degraded sal coppice forests, in the districts of Midnapore, Bankura and Purulia. This accounts for almost 47 per cent of the forest area in these three districts. The FPCs collectively have about 150,000 participating members. They have been most successful where the forest to household ratio is high; that is, where the dependence on forests for livelihood security is the greatest. The functioning of the FPCs has also been facilitated by prompt action by the Forest Department on complaints by villagers regarding illegal harvesting by outsiders. While the regrowth of these predominantly sal forests has been impressive, other trees such as mahua, kusum, amla, neem and karanj have also benefited from villager–Forest Department protection and co-operation. Where villagers were earlier forced to look for work elsewhere, the cumulative benefits of joint forest management have resulted in a significant reduction in seasonal migration out of these areas. Restored to effective control over their environment, ecosystem people are no longer forced to become ecological refugees (Malhotra and Poffenberger 1989; Malhotra *et al.* 1991; West Bengal Forest Department 1988; Steward 1988).

We may highlight five key features of the success of the FPCs in southwestern Bengal:

(a) No additional funding has been required for these schemes. In fact, Forest Department employees have devoted considerably less time and effort to organizing FPCs than they previously devoted to policing. This has important implications for Forest Department–local community relations elsewhere in the country.

(b) The benefits of FPCs have cut across both ethnic and political boundaries. FPCs have been successfully formed in villages with tribal, non-tribal and mixed populations, as well as in villages owing allegiance to the Congress, Communist Party of India (Marxist) and Jharkhand Parties. Interestingly, a study of forty-two FPCs in one part of Midnapore district revealed that all-tribal FPCs and mixed FPCs with a higher proportion of tribals were performing best of all – this being ascribed to higher tribal dependence on forest produce and tribals' more intimate knowledge of forest ecology (Malhotra *et al.* 1991).

(c) As the scheme has been successfully extended to more than a thousand villages, it cannot be said to be an isolated case or a flash in the pan. Rather, it has been tried and successfully tested in a variety of situations.

(d) The experience of the FPCs has further undermined the conventional wisdom, on which much forest management has been hitherto based, that timber is the main produce of the forests. Thus the regrowth of sal has facilitated the emergence of a large diversity of plants in the understorey that have an incredible variety of local uses as food, fuel, fodder, medicinal plants and sundry processing materials for household use and sale. One study documents 155 different species of plants and animals as being harvested by villagers from sal forests (Malhotra *et al.* 1991). Of these, a flora of at least seventy plant species were identified as being used regularly and in substantial quantities. These included sal seeds and leaves, mahua flowers and seeds, kendu leaves, tubers, mushrooms and cocoons. On a conservative estimate, 17 per cent of household income, on the average, came from the collection of non-timber forest produce. Estimated over a ten-year period, this income would be seven times more than the realization of the 25 per cent share of the harvest of sal poles which the Forest Department initially offered villagers as the main economic benefit of joint forest management.

(e) Finally, and more than anything else, the experience of the FPCs points to a qualitatively new relationship between the Forest Department and local people. While the West Bengal Forest Department is still in effective control of these areas, it has shown a greater willingness than its counterparts in other states to share power, authority and economic benefits with the villagers. At the same time, the successful functioning of the FPCs has restored a sense of dignity and self-worth among these communities, now confirmed as joint managers of the forest.

Two independent but close observers of the evolution of joint forest management offer this evaluation of its progress:

> The lesson we draw from the forest protection committees of West Bengal is one of hope for the state, for the nation, and for the world. The process of forest degeneration can be reversed. Through partnerships between foresters and forest communities, effective protection can be established with ecological and economic benefits for the community and the larger society. However, the task is not an easy one. It requires the political will of the state to delegate responsibilities to the forest communities, and changes in policies and procedures that may have been in effect for over a century. Ultimately, it requires a transition from management practices developed during the 19th century, to a management system that can respond to the social, economic and ecological needs of the 21st century. This requires dynamic leadership within the forest department to allow for a transition from traditional modes of timber production and forest policing, to a cooperative, responsive ability to work with rural communities.
>
> (Malhotra and Poffenberger 1989)

The experience of FPCs in West Bengal over the past two decades offers a fascinating comparison to the rather longer history of *van panchayats* in the Uttar Pradesh Himalaya. There are some major differences between the two schemes. As regards the respective social structures, hill villages are marked by comparatively little differentiation, while the villages of West Bengal are characterized by considerable heterogeneity with respect to caste, class and ethnic group. From an economic point of view, Himalayan forests are valuable to local communities in large part because of the close integration of agriculture and animal husbandry – with grass in pine forests and oak leaves as fodder being particularly prized. By contrast, the tribe and caste groups of the West Bengal FPCs cherish a much greater variety of non-timber forest produce, collected both for consumption and sale. Finally, *van panchayats* have to contend with an indifferent and even hostile official environment: nowhere are bureaucrats and politicians as authoritarian and corrupt as in Uttar Pradesh, while in recent years the state has tried hard to bring *van panchayats* more closely under its control. The FPCs of West Bengal have the inestimable advantage of functioning in a more congenial environment – with sympathetic politicians committed to decentralization and rural development, and bureaucrats providing inspirational leadership in the spread and functioning of protection committees.

These divergences apart, there are some notable similarities as well. In both cases, forests are in relatively good condition – ecologically speaking, certainly in better shape than under previous management systems. Common, too, is the diminution of conflict between peasants and the state over forest resources – though in the one case (West Bengal) this has been achieved through genuine partnership, in the other (Kumaun) by marking out distinct, non-overlapping areas for the Forest Department and village communities to manage. Third, in both schemes women are among the main beneficiaries of forest management. As primary gatherers and collectors of forest produce, women stand to gain most from forest protection and regeneration. Admittedly, it is only in the exceptional case – for example, where motivated voluntary organizations are active (Burra 1991) – that women have a formal leadership role in *van panchayats* or forest protection committees. But they do play a key informal role in the detection of intruders and those who violate the rules; in the forest much of the time, they are in effect the eyes and ears of the village community. Fourth, both schemes are genuinely decentralized, with benefits flowing largely to the village instead of (as is always the case with commercial forestry) to more powerful omnivore groups outside.

In their own, very different, ways, *van panchayats* and FPCs offer alternative paradigms to the authoritarian system of state forestry that has prevailed over much of India for well over a century. Where the latter rests on the exclusion (through policing) of rural communities, these systems work on the principle of inclusion. By restoring local control over management and use, they have overcome, with some success, the alienation of villagers from forests that has almost everywhere been the unhappy consequence of state forestry.

Notably, neither system has exhausted its potential. Thus *van panchayats* constitute just over 7 per cent of the forest area of the districts in which they operate – and there is no reason why, as concerned local scientists and activists have demanded, the large areas of 'civil and soyam' forests, presently open-access, heavily degraded forests under the nominal control of the district administration, should not be brought under the more effective and equitable *panchayat* system (see Shah 1989). In other areas with comparable, relatively undifferentiated, social structures, the constitution of a network of village forests, with the administration playing an advisory and supporting role, might be a worthy aim of state policy. In the majority of forest regions, however, it is more likely that a system based on the West Bengal experience of joint village community–Forest Department management is able to succeed. Sadly, despite the repeated urgings of sympathetic officials in the central Ministry of Environment and Forests, and although some individual forest officials have, on their own, started FPCs in their division, other state forest departments are lagging far behind in this regard. Political corruption, the territorial instincts of forest officials and the sloth and inertia of Indian administrations generally constitute formidable obstacles to the spread of joint forest management outside the state of West Bengal. Yet these obstacles must be overcome if India's forests are to be managed equitably, sustainably and without friction between state and citizen.

CONCLUSION

Let us now recapitulate our suggestions. The structure for the administration of public (including forest) lands remains essentially colonial in nature. While reform of agricultural land was pressed forward following independence, the management of public lands has remained frozen. Obviously it too needs a radical reorientation. These lands should be divided into two categories: (a) lands devoted to ecological security, and (b) community-managed lands devoted to providing livelihood security through a production system compatible with ecological security. The commercial plant production function should be fully shifted to private agricultural lands. Given such an outlook, the foresters would play the role of joint managers with people of lands devoted to ecological security or to livelihood security and an extension machinery serving tree farmers. Notably, the assignation in the past of overlapping, often conflicting, functions to a unitary forest department as well as the same patch of forest, without a clear priority being assigned among these functions, has in many areas been a prime factor behind deforestation. Therefore, it would be best, as we suggest, to separate these functions. This management system should be worked out and implemented on the basis of a detailed decentralized land use planning exercise which would start afresh with land capability rather than the nature of bureaucratic control of land as its starting-point. Once an appropriate land use plan, with emphasis on the urgency of ecological security and livelihood security, is worked out, then its proper implementation could be organized not as a centralized bureaucratic exercise but as a location-specific, people-oriented exercise. This calls for strengthening of the village- and district-level planning and administration machinery with higher-level controls primarily geared to ensure that the twin considerations of ecological security and livelihood security are given due weight.

This separation of objectives, functions and management systems for the three main categories of land use outlined above – nature reserves, community forests and farm forests respectively – must be the starting-point of forest administration. A shift away from state monopoly is an essential precondition for both ecological security and livelihood security. Indeed, active involvement of the people is also necessary for alleviating the bitter, wasteful and often violent conflicts between the state machinery and the rural population over access to forest produce. That healthy forest cover can be brought about only through a close co-operation between government and the villagers was well realized by one of our early nationalist organizations, the Pune Sarvajanik Sabha. Contesting the colonial Forest Act of 1878 for its excessive reliance on state control, the Sabha pointed out that the maintenance of forest cover could more easily be brought about by

> taking the Indian villagers into confidence of the Indian government. If the villagers be rewarded and commended for conserving their patches of forest lands, or for making plantations on the same, instead of ejecting them from the forest lands which they possess, or in which they are interested, emulation might be evoked between neighbouring villages. Thus more effective conservation and development of forests in India might be secured, and when the villagers have their own patches of forests to attend to government forests might not be molested. Thus the interests of the villagers as well as the government can be secured without causing any unnecessary irritation in the minds of the masses of the Indian population.

More than a century on, these sentiments remain strikingly relevant. For we are yet in search of a truly democratic and participatory system of forest management; one founded not on mutual antagonism, but on a genuine partnership between the state and its citizens.

8

IS THERE SAFETY IN NUMBERS?

NO MORE INDIAS?

A decade after India attained independence, the writer Aldous Huxley was deeply pessimistic about the future of a culture he had closely studied and long admired. As he wrote to a friend:

> India is almost infinitely depressing; for there seems to be no solution to its problems in any way that any of us [would] regard as acceptable, the prospect of overpopulation, underemployment, growing unrest, social breakdown, followed, I suppose, by the imposition of a military or communist dictatorship.
>
> (Grover Smith 1969: 926)

It is noteworthy that Huxley puts 'overpopulation' at the top on his list of Indian problems (in another letter from that 1961 trip, he wrote to his brother, the biologist Julian Huxley, of 'the impossibility of [India] keeping up with the population increase'). Indeed, he implies that there is a direct causal link between growing numbers and the other problems he alludes to. When the environmental debate acquired force in the West a few years later, one of the most vocal strands in the debate likewise held overpopulation to be the prime reason for ecological degradation. Inevitably, the discussion focused heavily on India, which came to be regarded by many Western environmentalists as the classic basket case, unable adequately to feed, clothe or house its growing population.

Modern environmentalists who focus on the 'population problem' are usually termed 'neo-Malthusians' after the English parson, Thomas Malthus, who in the late eighteenth century first predicted that the growth in human numbers would outstrip the growth in food supply. Indisputably the best-known neo-Malthusian is the Stanford biologist, Paul Ehrlich. This is how Ehrlich begins Chapter 1 of his 1969 book, *The Population Bomb*:

> I have understood the population explosion intellectually for a long time. I came to understand it emotionally one stinking hot night in Delhi a couple of years ago. My wife and daughter and I were returning

> to our hotel in an ancient taxi. The seats were hopping with fleas. The only functional gear was third. As we crawled through the city, we entered a crowded slum area. The temperature was well over 100, and the air was a haze of dust and smoke. The streets seemed alive with people. People eating, people washing, people sleeping. People visiting, people arguing and screaming. People thrusting their hands through the taxi window, begging. People defecating and urinating. People clinging to buses. People herding animals. People, people, people, people.
>
> (Ehrlich 1969: 15)

In this highly charged description, exploding numbers are held guilty of pollution, stinking hot air, and even technological obsolescence (the 'ancient taxi'). If the Malthusians are right we might as well tear up our wills and wait for doomsday. In the grim scenario painted by them, purposive human action might almost seem pointless – for no development strategy, whether 'business-as-usual' or conservative-liberal-socialism, could succeed in the face of exploding numbers. But of course the case of the Malthusian ecologists, whose motto is 'No More Indias', is flawed in many respects. For one thing, it relies heavily on a mistaken analogy taken from the living world. This is the concept of 'carrying capacity', used by biologists to define the maximum number of individuals of a species that can be supported by a particular habitat. But when applied to human populations, the concept is a slippery one. It fails to take account of the ingenuity of people, their abilities through technical change to squeeze more out of a given habitat. Moreover, the needs of humans, as distinct from animals, are culturally determined, while (again unlike animals) they might have access through transport and communication to resources from far-flung localities (see Hartman 1987).

In the West itself, the arguments of the neo-Malthusians were quickly subject to critical scrutiny. From the right, liberal economists accused them of gravely underestimating both human ingenuity and the abundance of natural resources: one scholar even claimed that as intelligent and creative individuals were the 'ultimate resource', rapid population growth was in fact a good thing (Simon 1981). (These claims draw strength from countries such as the Netherlands, with a higher population density than Bangladesh but hugely prosperous none the less.) From the left, socialists disputed the contention of Malthusians that growing numbers were the prime cause of environmental degradation in the West, arguing instead that it was imperfect technology and the workings of the market that were responsible for pollution and resource exhaustion (see Commoner 1971). The Malthusians have also been taken to task for their apparent preference for coercive measures to bring about family planning and birth control.

Not surprisingly, within India left-leaning environmentalists have been quick to take offence at Malthusian prognoses, the more so as India's own indigenous omnivores, when called upon to explain environmental degradation or social strife, invariably point an accusing finger at the propensity of ecosystem people to breed in large numbers. Most commonly, the rebuttal by environmentalists of these arguments invokes the question of consumption. Whether judged in energy or material terms, an average American's demands on the earth's resources are at least an order of magnitude greater than those of an average ecosystem dweller of the Third World. The birth of one American child will thus have an environmental impact equal to that of (say) the birth of several dozen Bangladeshi children. By this reckoning, if there is a population problem at all, it exists in affluent consumer societies such as the United States.

This is a line of argument that is difficult to dispute, but some Indian environmentalists have gone on to dismiss altogether the role of population in ecological degradation. Our own line of analysis also suggests that there are more important structural factors behind the massive degradation of India's land and water resources. But surely it is going too far to reject the need for family planning, as the otherwise estimable second citizens' report does while claiming that 'If India's people were to go hungry, it can be said with authority that it would not have *anything* to do with their number but with the callous mismanagement of the country's natural resources' (Centre for Science and Environment 1985: 162; emphasis added).

Where one school, characteristically Western in origin, thus regards population growth as the main contributory factor to environmental abuse, another school, quite dominant in India, tends to treat the question of human numbers as irrelevant to the development debate. Both positions are obviously too extreme. Without in any way underestimating the part played by inequity and inefficiency in causing environmental degradation, one must recognize that in many local situations rapid population growth is indeed exerting unsustainable pressures on grazing land, water resources or food supply (Jodha 1986). Yet the way out does not lie in either of the two solutions favoured by omnivores, global and national, namely, coercive family planning, or leaving the ecosystem people (who are primarily responsible for population growth) to their own fate. What then is the outlook of the conservative-liberal-socialist in relation to the population–environment nexus?

We believe that people are motivated above all by their perceived self-interest and the interest of their family members, or to a lesser degree that of some small, homogeneous social group to which they belong. They are rarely moved by the interest of a large heterogeneous group, especially if that group be characterized by high levels of inequality. Human populations will, then, tend to grow when people perceive their interests to lie in producing several offspring. Such is likely to have been the case over much of human history. After all, the entire present human population probably derives from a parental population of a few thousand *Homo sapiens sapiens* who might have developed the current linguistic abilities around fifty thousand years ago (Cavalli-Sforza *et al.* 1994). On this assumption human populations have on average been doubling every 350 years. With mortality rates reported for pre-industrial populations of Europe, an average woman reaching adulthood would need to bear slightly over six offspring for populations to grow at this rate. Evolution seems to have led all animals to tend to want to rear a goodish number of offspring; it is quite plausible that humans too tend to identify self-interest with bearing six or seven children (Dawkins 1976).

Human population growth may, then, be decelerated only when humans come to view their self-interest as consisting in a much smaller number of offspring; or when the mortality rate is particularly high owing to war, disease or starvation. Leaving out the second possibility, there seem to be three scenarios under which humans might voluntarily produce a smaller number of progeny:

(a) Among highly mobile hunter-gatherers women might find it impossible to be on the move carrying more than one offspring at a time. If so, they must pace children about four years apart, perhaps by suckling each child over a long period.
(b) Among highly sedentary hunter-gatherer-horticulturists with limited territories and in acute conflict with neighbouring tribes, groups may be at a disadvantage when growing too large in size, exceeding the carrying capacity of the territory and occasionally facing severe food shortages. They may therefore restrain population growth in the interest of their relatively small, egalitarian groups.
(c) Among modern industrial societies the young may be obliged to spend a long time acquiring the necessary training to handle the complex artefacts and information necessary for gainful and status-worthy employment. Under these circumstances the young cannot help their parents by adding to the family income; instead, parents have to invest heavily in enhancing the quality of their offspring. The parents may then identify their self-interest, as well as that of the offspring, in producing a small number, perhaps one or two of them (Caldwell 1982).

The world over, human societies are rapidly moving away from the first two scenarios. A number of industrial countries have, however, undergone a demographic transition over the past century corresponding to the third scenario. In India too there is evidence of such a transition taking place among certain strata of the population. Perhaps the best set of such evidence comes from the People of India Project of the Anthropological Survey of India (K.S. Singh 1992). This project attempts to map the entire human surface of India based on interviews with over 25,000 individuals belonging to the 4,635 communities to which the entire population of the country has been assigned. These interviews, in conjunction with the collation of other available information, permit each community to be characterized by some 500 traits relating to ecology, food habits, social organization, occupation, social and economic status, and response to modernization. Information has also been gathered on the number of offspring perceived to be desired by most members of any given community – recorded as 1 or 2, 3, 4 or more. The data thus lend themselves well to an analysis of how a variety of traits characterizing any particular community relate to the number of offspring desired by members of that community.

The data clearly show that communities that report a desire for a small number of offspring display the following features:

(a) They are engaged in occupations in the organized industry–services sectors, or are owners of land under intensive agriculture.
(b) They tend to have access to modern amenities such as electricity and tap water, television and agrochemicals.
(c) Both boys and girls in the community receive high levels of education.
(d) Most of the communities are urban, or have combined rural–urban distribution.
(e) Most of the communities involved belong to the upper-caste groups.

This is a picture consistent with the third scenario we outlined above, that of the demographic transition in industrial societies. In such communities parents have an interest in imparting high levels of education to the offspring over many years. Indeed, to have both boys and girls studying up to postgraduate level – that is, to the age of 22 or more – is a feature of many communities wanting few children. This education is essential in equipping them to find opportunities in the modern sector, a sector involved in the use of complex artefacts that require extensive learning for effective handling. Since activities in the modern sector are largely concentrated in urban localities, and towns and cities have been provided with modern amenities, it is to be expected that most such communities would have access to these amenities. It is the upper castes that dominate the modern industries–services sector for a variety of historical reasons, and it is these who have largely come to desire few children.

Communities desiring four or more offspring on the other hand exhibit a contrasting set of attributes:

(a) They are engaged in occupations primarily involving agricultural labour, herding, hunting-gathering or fishing.
(b) They have limited access to modern amenities. In most such communities women must fetch water and fuelwood, and help in fishing or looking after livestock.
(c) Neither boys nor girls in these communities receive much education; many drop out at an early age to help out their families.
(d) Children in many such communities are engaged in paid labour.
(e) Numerous such communities are becoming progressively impoverished; they are largely landless; they indicate recent involvement in or increased

dependence on wage labour, and decreased consumption of luxury food such as fruit.

(f) Many such communities are from tribal or lower backward castes; all are exclusively rural or forest dwellers.

For these communities there are no opportunities to invest in children's education to equip them to enter the organized industries–services sector. Since both men and women are engaged throughout life in unskilled labour, children can very quickly achieve the status of earning members of the family. They thus become an economic asset at an early age and members of these otherwise mostly assetless communities are motivated to produce a large number of offspring. Notably enough these are also communities most directly dependent on natural resources such as fuelwood. Their continually increasing population does imply an ever increasing pressure on the dwindling stock of the country's natural resources.

One may visualize five routes towards a deceleration of population growth:

(a) Compulsion to limit the number of offspring, practised with some success in China, but which turned out to be a disastrous failure in India during the Emergency years of 1975–7.
(b) Provision of specific, one-time incentives, such as cash rewards for undergoing vasectomies or tubectomies.
(c) Exhortation, advertisements on television and so on.
(d) Improved access to information and the hardware of family planning.
(e) A restructuring of the economy and the society leading to widespread motivation to produce a smaller number of offspring.

It is our contention that the first three routes can have only limited positive impact; indeed, the attempts at coercion during the Emergency were greatly counterproductive in the long run. Improving access to information and drugs, implements or surgical operations for family planning is undoubtedly of value. But in itself this will not be enough. Indeed, the Anthropological Survey of India data reveal that in all the states and union territories of India many communities employ modern family planning methods; but only a fraction of these favour one or two children. A significant number of communities employing modern family planning methods still desire three, four or more children. Unless there is a change in motivation, access to family planning methods by itself will not work.

What form of restructuring of society and the economy will motivate most of India's people to produce a small number of offspring? We believe it to be neither the ongoing inequitable, resource-exhausting pattern of development, nor the return to an agrarian-pastoral society favoured by a section of environmentalists. Instead, the path of equitable and resource-efficient development advocated by us would be most compatible with finally bringing the growth of India's population to a halt. The key factor is the value attached by society to skilled manpower – skills that require years of patient investment by parents and specialist teachers. The skills required for low-input subsistence agriculture and animal husbandry are rapidly absorbed at home by boys and girls; parents can therefore produce large numbers of them without interfering with their training. Indeed, the traditional blessing to a newly married girl in India, 'bear eight sons over a long married life', represents the value conferred on large families by an agrarian society.

Large investments in the training of offspring are called for when they must learn to handle complex artefacts: machinery, chemicals and sophisticated technical information. The industrial societies of the United States, Europe and Japan have created conditions under which the majority of people have jobs in either industry, services or highly mechanized/chemicalized agriculture requiring such training. But this is achieved through investing large amounts of energy, material and informational resources to build up these sectors. That is why per capita consumption of resources in these countries is more than ten times that in countries like India (Durning 1992). India simply does not have at its disposal sufficient levels of energy or material resources to support the employment of every one of its citizens in the modern sector. This is in part why organized industry has failed to create a demand for trained personnel at rates that could substantially reduce the population pressure in the Indian countryside. Irrigated, intensive agriculture has done better, and substantial new job opportunities have opened up, although mostly for unskilled labour, in pockets of productive agriculture. But resources have been used extremely inefficiently in both industry and intensive agriculture, so that resource investments have yielded very inadequate returns in terms of employment for skilled personnel in both these sectors. In part this has been compensated for by the explosion of job opportunities in the services sector – especially in governmental and quasi-governmental organizations. But this is counterproductive because this sector, above all, promotes wastage of resources. Faced with a limited resource base given its large population, and quite unable to bring in resources from outside, India has further failed to use what it has in an effective fashion to create new job opportunities for trained personnel. This is the root cause of India's failure to halt the explosion of its large population.

The only way, then, is to try to make the most of India's natural resources, to conserve them, to use them ever more effectively. Obviously Indians must come to adopt the Japanese philosophy of patiently getting more and more out of the natural resource base, year in and year out, without of course emulating Japan's appetite for natural resources from outside its borders. This will go hand in hand with a broader-based development, doing away with the current pattern of enclaves of industry and intensive agriculture prospering parasitically by guzzling on the resources of hinterlands that in turn become progressively impoverished. This will be a strategy of many small, timely interventions, focusing on a rebuilding, not exhaustion, of the natural resource base. Such a process would be put in place only with local communities being handed back control over the local resource base, sustainably developing these resources using local labour as well as folk knowledge and wisdom suitably married to modern scientific understanding. Only such a process can enhance the quality of life of the masses of the Indian population, lifting them out of the compulsion to liquidate whatever resources they can lay hands on to eke out a living. Only such a process has the potential to create job opportunities for large numbers of trained people to participate in a process of carefully planned development of the local natural resource base. Only such a process holds the promise of motivating the Indian people on a broad enough scale to bring the ongoing explosion of population eventually under control.

9

RESOURCES OF HOPE

THINK GLOBALLY, ACT LOCALLY

Ecologists, the American conservationist Michael Soule has remarked, live in a world of wounds: wounds to both nature and society, as witness the intensification of social conflict that has almost everywhere accompanied the devastation of the Indian countryside. In this book, we have documented the range of factors behind the tearing up of the social and natural fabric of India. But we have also tried to document the myriad attempts to heal these wounds, the quiet, persistent and often unhonoured struggles to mitigate social inequalities and restore the health of the environment.

Ours is a counsel not of despair but of hope, and it is with a recapitulation of these positive initiatives that we wish to end our narrative. We focus first of all on the district of Uttara Kannada in the southern state of Karnataka – a district that captures in microcosm many of the processes of ecological change described in this study. This is a region of low, undulating hills that run right into the Arabian Sea to the west and merge with the Deccan plateau to the east. The hills are extensively wooded, with over 60 per cent of the land under forest cover. The region has excellent rainfall, with annual precipitation exceeding 5,000 mm near the crestline of the Western Ghats. From this tract of heavy rainfall originate short west-flowing rivers that descend steeply to the coastline, as well as tributaries of the great river Krishna which flows eastwards. The district has a relatively low density of human population at 104 persons per sq km in 1991. Fishing, rice, coconut, betelnut, pepper, cardamom and cotton cultivation and manganese mining are the mainstay of the economy. The district also supports fish processing, paper, plywood, tiles and caustic soda industries. The people of the district have traditionally depended heavily on the forests for fuelwood, grazing, leaf manure, construction material and a wide range of non-wood forest produce.

Hand in hand with this intimate link with forest resources go numerous traditions of conservation and sustainable use. As illustrated in Chapter 2, despite official hostility some villages of Uttara Kannada have managed successfully to protect and manage community forests that through regulated harvests continue to meet their requirements of fuel, fodder and small timber. Other patches of forest are preserved as sacred groves, and given total protection by peasants. One of the most notable of these is the grove of Karikanamma (= mother goddess of the dark forest), a magnificent stand of dipterocarp forest perched on a high hill looking far out to the Arabian Sea. The Western Ghats have two species of *Dipterocarpus*, the flagship genus of the tropical rain forests of Asia. Uttara Kannada is the northernmost limit of the geographical range of this genus on the Western Ghats; and *Dipterocarpus* has nearly vanished from everywhere else in the district. It has disappeared elsewhere because it was greatly valued by and overexploited for the plywood industry. It persists in the grove because the long-held beliefs of ecosystem people have prevented its felling in this one sacred forest.

Studies suggest that as much as 6 per cent of the land in Uttara Kannada was once under sacred habitats: groves, ponds, pools in river courses. Today this proportion has come down to just 0.3 per cent in a 5 km x 5 km area we studied intensively in the Siddapur *taluk* of the district. This area still retains fifty-two sacred groves, some as large as 2–3 ha scattered throughout the countryside. These groves currently harbour remnant patches of evergreen rain forest, refuges for many species that have become scarce over the rest of the district. The lofty trees represent large amounts of cash, cash that cannot but be a great temptation to the poor peasants who are continuing to protect the groves, sometimes to the extent of not even removing fallen fruit. Foresters have been quite unmindful of the value of these traditional systems of conservation till recent times; indeed, some groves of over 100 ha in size in this district were clear-felled by the Forest Department to raise eucalyptus plantations in the 1970s. There are pressures on the groves from some segments of the village community as well, from people who are hungry for cash or land for cultivation. Yet recent years have seen a growing recognition of the significance of these wholly indigenous systems of conserving biodiversity – and scientists, social activists and environmental organizations have all joined hands in studying these practices and campaigned for their continuation (Subash Chandran and Gadgil 1993).

Whereas sacred groves testify to the persistence of age-old conservation practices, the district of Uttara Kannada has also contributed substantially to the modern environmental movement, beginning with the opposition to the Bedthi dam project in 1979–80. The Bedthi is one of the four major west-flowing rivers of Uttara Kannada, the others being the Kali, Aghanashini and Sharavathy. These rivers originate in the Western Ghats at altitudes of around 600 m and drop abruptly to sea level. Just on the southern border of Uttara Kannada, in the district of Shimoga, are the Sharavathy waterfalls. The story goes that Sir M. Visvesvarayya, the famous engineer-statesman, exclaimed, 'Oh, what a waste of energy!' when he first saw these falls. The waters of the Sharavathy were duly harnessed through two hydroelectric projects in the early years of independence. This was followed by a series of dams on the

Kali initiated in the 1960s. Plans were then set in motion to tap the hydroelectric power of the Bedthi and Aghanashini.

Environmental impact assessment of major projects became mandatory in 1978, and a hydroelectric project on the Bedthi river came up for review. One of the present writers was a member of a committee set up to prepare such an assessment. The committee met one morning in the state capital of Bangalore, was briefed by the state power corporation and then drove the 500 km to the Bedthi valley. We reached the actual project location next morning, spent just three hours touring around the dam site and a small part of the submersion area and then sat down to write the report. This was clearly being treated as a formality to be got over, with no serious intention of any careful assessment of the environmental impact of the project. Indeed, the region was quite unfamiliar to the other committee members. But a report recommending that the project be awarded an environmental clearance was pushed through despite minority protest.

But a good proportion of the farmers whose lands were slated for submergence in the Bedthi project were well-educated betelnut, pepper and cardamom producers belonging to the upper-caste group of *Haviks*. Roused against this charade of an environmental clearance, their well-organized co-operative society funded an alternative assessment. This was helped along by the fact that a dissenting member of the committee could pass on to them the basic information on project design, information that is normally withheld from all ordinary citizens. This alternative assessment showed that the original official exercise was seriously flawed, for it deliberately under-estimated adverse impacts. Not only were a variety of environmental con-sequences ignored, but proper computations suggested that the economic benefit : cost ratio was less than 1.5, the standard set by the Planning Commission for acceptance of a power project. This fresh assessment was followed by an open seminar presided over by the doyen of Kannada literature, Dr Shivaram Karanth, and addressed by, among others, the *Chipko andolan* leaders Sunderlal Bahuguna and Chandiprasad Bhatt, and by Professor V.K. Damodaran, a stalwart of the Kerala Sastra Sahitya Parishat. Officials of the Karnataka government's Forest and Planning Departments also participated, although the Power Corporation pulled out at the last moment. Held in January 1980, this was the first-ever public hearing on a development project in India. After this, the project was shelved (Sharma and Sharma 1981).

In his valedictory address to the Bedthi seminar, Dr Shivaram Karanth had stressed that the betelnut gardeners too were guilty of the wasteful use of natural resources. Some sensitive individuals among them rose to the challenge of setting their own house in order and through a co-operative society launched a project for the good management of the soil, water, vegetation and livestock in and around their villages. Beginning in the early 1980s the project brought together farmers, scientists from agricultural and technical institutions, voluntary agencies and government officials in a co-operative effort. Around the same time the government of Karnataka had launched its own progressive programme of integrated development of watersheds, bringing together specialists from its various agencies under the umbrella of a dryland development board. This programme also visualized involvement of local villagers as partners in the good management of private and public lands. The programme, initially focusing on the drier areas of the state, provided an appropriate framework for carrying forward the project initiated by the betelnut gardeners of Uttara Kannada.

Over a decade now the betelnut gardeners have been pursuing attempts at an integrated development of land, water, vegetation and livestock in their hilly terrain. At times there have been frustrations with different interests among the village population pulling in different directions, as well as irritation with the operation of the government machinery that has been funding these efforts. But there have been some most encouraging successes as well. Two of these include development of fodder resources and diffusion of fuel-efficient wood stoves. The youth club in the village of Bellikeri in Sirsi *taluk*, for example, has organized a co-operative fodder farm on forest land assigned to one of its members. Gradually the production of fodder is becoming self-supporting on the basis of the sale of hand-harvested fodder grass to the local villagers (Prasad *et al.* 1985).

Another successful experiment has involved propagation of fuel-efficient wood stoves based on a design developed by highly qualified chemical engineers of the Indian Institute of Science in Bangalore. In the rest of the state this programme, dependent on government subsidies, has run into problems, but among the betelnut gardeners there is real concern about the erosion of forest resources in their vicinity. The horticulturists are also interested in using newer sources of fuel such as betelnut husk. The new design of their bathwater heating stove permits the use of this rather abundant and otherwise unused resource as fuel. In consequence, new fuel-efficient wood stoves have diffused quite rapidly in the community.

Even as the farmers of the district were coming together for more responsible resource use, the Bedthi project was revived by the government of Karnataka in 1991. Expectedly, the revived project has met with opposition. Strikingly, this time protests have been accompanied by a bold alternative design offered by the people of Uttara Kannada. With the help of a retired engineer who once headed the Karnataka Power Corporation, the betelnut gardeners have been able to propose an alternative scheme to generate power, involving sixteen smaller reservoirs in place of the single large reservoir planned by the government. The sixteen smaller reservoirs would submerge much less land under forest, cultivation and human habitation, and are therefore an option more acceptable to the local people. This alternative design could in fact generate a larger total amount of hydroelectric power over the year, although the single large reservoir could produce more power during the dry season. More importantly, the sixteen-reservoir alternative would generate power at a substantially lower cost than the power corporation's design. The alternative design also visualizes possibilities of using tidal flow as well as irrigated energy plantations to further augment power production.

Competent engineers have been involved in working out this alternative design, and it is quite likely that their claims would hold up under close scrutiny. What is crucial here is that the government monopoly is being broken – and through an initiative of the local people. Indeed, the betelnut gardeners are exploring the possibility of floating a public company to execute the project, which they are confident is economically far more viable than the State Power Corporation's project. At all levels, then, from sacred groves and village forests to major hydroelectric projects the resourceful people of Uttara Kannada are giving us hope that the tide of environmental degradation might yet be turned around; and in ways that would benefit the broader masses of the population.

A TRADITION TO LIVE UP TO

These initiatives in Uttara Kannada exemplify a strain of constructive social activism and critical enquiry that runs deep in Indian culture. We might thus situate these initiatives in the context of the traditions upheld by the remarkable individuals to whom we have dedicated this book. Like Shivaram Karanth (the grand old man of the environmental movement in Uttara Kannada), the great ornithologist Salim Ali was tirelessly active well into his eighties, mapping India's rich biological diversity. As one who was always alert to the need to integrate conservation with local needs and traditions (see Ali 1977), Salim Ali would have greatly approved of the efforts under way to protect sacred groves. The social reformer Jotiba Phule, himself from the stock of ecosystem people, was one of the first to bring to wider notice the close integration of agriculture and forestry, a link he saw being broken by the takeover of forests and common lands by the state (see Phule 1883). The rehabilitation of the social and natural world of rural India was also the lifelong concern of those two great Gandhians, J.C. Kumarappa and Mira Behn. It was Kumarappa (1946) who first wrote of the 'economy of permanence', and both he and Mira Behn spent many years understanding the workings of Indian agriculture and trying to restore it to a sound ecological footing.

Outside Uttara Kannada, indeed in almost all other parts of India, social activists are upholding the traditions associated with those such as Phule, Kumarappa, Mira Behn and Salim Ali. Some of these initiatives focus more on resistance to environmental destruction or resource capture; others on ecological restoration and the adoption of environment-friendly technologies. Some have tried to adapt modern technology to the rural context;

others have sought instead to draw upon and revive the rich traditions of prudent use among ecosystem people – such as community wood-lots, sacred groves or rain-fed tanks. These efforts have been guided by Gandhians, Marxists and wholly apolitical social workers. Some initiatives, like the work of Chandi Prasad Bhatt and the Chipko movement, or the Baliraja dam in drought-prone Sangli, have been extensively written about; others are unknown outside their immediate locality, though none the less important for that.

There has even been the odd initiative emerging from the omnivore sector. Thus the Tatas, long the most progressive of India's industrial houses, were found in a recent study of magnesite mining in the Himalaya to be exceptional in their concern for minimizing the environmental impact of opencast mining, while at the same time providing infrastructure and employment to local villagers (Institute of Social Studies Trust 1991). Nor have the communists been lagging in this regard. Thus one of the most remarkable environmental success stories of recent years comes from the Marxist-ruled state of West Bengal, the village forest protection committees studied in Chapter 7.

With regard to this particular initiative, there is little doubt that the West Bengal Forest Department has been far more willing than its counterparts in other states to share power with ecosystem people. Here the success of the village forest committees cannot be isolated from the wider processes of political change in the state. Since coming to power in 1977, the alliance of left-wing parties which has ruled West Bengal has crafted local level institutions that have real decision-making power, real control over financial resources. Moreover, the elected representatives at village and district levels have real authority over government officials. This decentralization of power has been accompanied by a substantial degree of land reform, so that rural society in Bengal is less inequitable than in the rest of the Gangetic plain (see Kohli 1987). Rising expectations of ecosystem people, and growing powers to enforce these expectations, have been coupled here with a move towards a less class-ridden society. This is the context in which local-level political leadership, and a more responsive bureaucracy, have worked together to bring about a broad-based and markedly successful programme of natural resource management in the sal forests of Bengal.

The forest protection committees of West Bengal are a perfect vindication of the core message of this book, the need to blend ecology with equity. They provide further testimony of the need for a genuinely decentralized political system country-wide, where powers to use natural resources lie not with insensitive and corrupt bureaucracies but with the people who most deeply depend on these resources. Today those in power, the omnivores of India, can successfully pass on the costs of resource abuse and environmental degradation to the masses of ecosystem people, and to ecological refugees. So long as this situation persists there is little real hope. But the self-interest of India's ecosystem people is congruent for the most part with the good husbanding of natural resources, at least in their own localities. The real solution for the long-term health of the environment thus lies in passing effective political power to the people.

The state of Karnataka, of which Uttara Kannada is part, itself flirted briefly with decentralized institutions on the West Bengal pattern between 1986 and 1990. A prime mover behind the constitution of *mandal panchayats* and *zilla parishats* in the state was the late socialist leader Abdul Nazir Saab, one of the most outstanding politicians in the history of free India, and chronologically the last of the exemplars to whom this book is dedicated. As Rural Development Minister, Nazir Saab was able to motivate a lethargic bureaucracy enough to provide drinking water to every village in the state, an act which earned him the appellation, richly deserved, of 'Neer Saab' (the man who brought water). Despite the fierce antipathy of the same bureaucracy, Nazir Saab and his colleagues in the Janata Dal Party – which came to power in Karnataka in 1983, after long years of Congress rule – were able to push through the *panchayati raj* scheme. In its brief life, the system underlined one cardinal truth: namely, that the interest of politicians in environmental issues is, by and large, inversely proportional to the size of their constituency. In our experience, Members of Parliament appear to care the least, members of state legislatures a little more, members of district councils a little more still, and members of *mandal panchayats* emphatically the most. A quarter of the seats at the two lower levels were reserved for women, who were often most vocal in calling, for example, for people-oriented forest management. Yet the system was not to the liking of omnivores, whose own interests it seriously threatened. When Congress returned to power in Karnataka in 1990, it moved quickly to suspend and eventually emasculate *zilla parishats* and *mandal panchayats*. Supporters of this move claimed that the decentralization of power only meant the decentralization of corruption. Yet the state Congress government that abolished the *panchayats* has since been acknowledged to have been the most corrupt administration ever to rule Karnataka. In December 1994 the Janata Dal returned to power in the state, promising to restore the *panchayat* system.

There is, of course, a real danger that an elite of upper-caste landlords would come to dominate lower-level political institutions. This fear prompted Dr B.R. Ambedkar, the great leader of India's lower castes, to oppose the devolution of power to village institutions at the time of national independence. He believed that only a disinterested elite, manning the agencies of the state, would be able to protect the vulnerable rural poor from exploitation by the dominant upper castes. This was also the hope of classical socialists, who likewise pinned their faith in the ability of the state to bring about a just and equitable society. But forty-five years of growing state power in India have painfully belied these hopes. The record of state-guided development is abysmal, when reckoned by the criteria of economic growth, environmental stability or social equity. It is this failure that lies behind our plea for a decentralized political system in preference to the continuing centralization of power in the hands of a narrow elite, the iron triangle of omnivores. But one must recognize that mere decentralization without redistribution would be self-defeating: hence the significance of land reform, education and health care, along with equitable access to common property resources. This would lead to the institution of a political system best summed up by the word 'empowerment', much as the colonial regime could be summed up by the word 'subjugation', and the system we presently live under as 'patronage'.

There are vast resources we can draw upon in moving towards such a political system. These include a rich history of social activism that stretches from Jotiba Phule to Abdul Nazir Saab and beyond, as well as the traditions of ecological prudence that still persist among ecosystem people in far-flung parts of India. In this book we have outlined a philosophical approach for bringing under one roof these varied traditions, recasting what may appear to be contesting ideologies in one unifying framework. Thus our emphasis on strong local communities borrows from the Gandhian tradition, that on democratic institutions and private enterprise from liberal capitalism, and that on equity from Marxism. This is an eclecticism that comes out of our reading of the ecological history of modern India, but we believe it also to be in the larger Indian tradition of tolerating, assimilating and synthesizing diverse strands. What we have called conservative-liberal-socialism might just provide the springboard for moving India towards a politics of empowerment, and towards an economy of permanence.

GLOSSARY OF WORDS IN INDIAN LANGUAGES

aankhbandi (allowance): literally, closing the eye. A bribe paid to an official not to take action against some violation.

andolan: a social movement.

ashram: traditionally abode of sages; currently used especially as a place to conduct constructive social activities.

astraole: a fuel-efficient wood stove designed by a group of engineers at the Indian Institute of Science in Bangalore.

ayurveda: Indian medicine.

baandh: dam.

babu: a white-collar worker.

bachao: save.

bandh: shutdown.

Bhoodan: literally: gift of land. A well-known social movement that urged landowners to gift land for redistribution favouring smallholders and landless peasants.

bidi: an Indian cheroot where tobacco is rolled in leaves of *Diospyros* trees.

chawls: multistoreyed tenements housing poorer families in urban areas.

Chipko: literally: to hug. A well-known environmental movement against forest destruction which started with peasants hugging trees to prevent their being cut.

dharna: a form of sit-down strike.

dhoti: a piece of cloth traditionally used as a lower garment by men.

gherao: a form of protest involving surrounding a person and preventing him/her from moving at will.

Gramdan: literally: gift of a village. A well-known social movement in which all landowners in a village gifted their lands for redistribution among the entire village population.

hartal: a strike involving closure of shops, factories, offices, transport facilities.

Jharkhand Mukti Morcha: a political movement for creation of a separate state of Jharkhand in tribal areas of central India.

jail bharo andolan: a form of protest involving courting arrest to fill up jails.

jal samadhi: a form of protest involving immolation in rising waters of a dam.

Jan Sangharsh: people's struggle.

Loot-mar-ka-mahina: literally: the month of plunder.

mandal: village cluster

mandal panchayat: literally: council of five members for a group of villages. In practice the council may include many more members.

mukti sangharsh: literally: a struggle for liberation.

nagarpalika: system of decentralized political institutions for the urban parts of India.

nallah: stream; *bandh*: stoppage.

Naxalites: groups working for armed revolution.

Naxalbari: a village in West Bengal where the activities of ultra-left Naxalites began.

padayatra: foot march; a form of public demonstration.

panchayat: a council of five members.

panchayati raj: system of decentralized political institutions for the rural parts of India involving *mandal panchayats* and *zilla parishats*.

pani panchayat: a council of villagers set up to manage common-property water resources.

panidari: a feudal system of ownership rights over a stretch of river.

parishat: assembly council.

pathashalas: schools in the traditional Indian system of education.

phad: a traditional system of irrigation in northern Maharashtra.

rasta roko: an agitation involving disruption of road traffic.

ryotwari: a system of assigning land where the rights of tillers were better recognized by the British authorities in southern and western India.

sangharsh samiti: struggle committee.

satyagraha: literally: insistence on truth. A form of non-violent protest popularized by Mahatma Gandhi.

shikar: hunting, particularly of bigger animals.

taluk: county.

targetbaji: unproductive pursuit of paper targets.

van panchayat: village forest council.

zamindari: a system of assigning land, introduced by the British authorities in eastern and northern India, where the local chieftains were made owners of very large tracts of land .

zilla: districts.

zilla parishat: an assembly of elected members governing a district.

GLOSSARY OF WORDS REFERRING TO INDIAN COMMUNITIES

Ambiga: a fisherfolk community of Uttara Kannada district in the state of Karnataka.

Badaga: a community of cultivators in the Nilgiris district of Tamil Nadu.

Brahman: a member of an upper-caste priestly community.

Halakki Vakkals: a community of lower-caste peasant cultivators in Uttara Kannada district in the state of Karnataka.

Haviks: a community of upper-caste priests-cum-horticulturists in coastal and hill areas of the state of Karnataka.

Kshatriya: a member of an upper-caste warrior community.

Sudra: a member of lower-caste peasant, herder or artisanal communities.

Vaisya: a member of an upper-caste trading/artisanal community.

BIBLIOGRAPHY

Achaya, K.T. (1993) *A Companion to Indian Food and Food Materials*, Delhi: Oxford University Press.

Agarwal, A. (1986) 'Human–nature interactions in a Third World country', *The Environmentalist*, 6.

—— and Narain, S. (1990) *Towards Green Villages*, New Delhi: Centre for Science and Environment.

Agnihotri, I. (1993) *Ecology, Land Use and Colonization: the Canal Colonies of Punjab*, mimeo, New Delhi: Nehru Memorial Museum and Library.

Ali, S. (1977) 'Wildlife conservation and the cultivator. Presidential letter', *Hornbill*, April–June, pp. 6, 36.

Alvares, C. (1989) 'No!', *The Illustrated Weekly of India*, 15 October.

—— (1992) *Science, Development and Violence: The Twilight of Modernity*, Delhi: Oxford University Press.

Anklesaria Aiyer, S. (1988) 'Narmada project: government–opposition alliance against Amte', *Indian Express*, 30 October.

Anon. (1991) 'The sangarsh yatra: a first hand account', *Narmada: A Campaign Newsletter*, Nos 7 and 8, April.

Areeparampil, M. (1987) 'The impact of Subarnarekha multipurpose project on the indigenous people of Singhbhum', in *People and Dams*, New Delhi: Society for Participatory Research in Asia.

Attwood, D.W. and Baviskar, B.S. (eds) (1988) *Who Shares?*, Delhi: Oxford University Press.

Bahuguna, S. (1983) *Walking with the Chipko Message*, Silyara (Tehri Garhwal district): Navjivan Ashram.

Bajaj, J.K. (1982) 'Green revolution: a historical perspective', *PPST Bulletin*, 2: 87–113.

Ballabh, N. and Singh, K. (1988) *Van (Forest) Panchayats in Uttar Pradesh Hills: A Critical Analysis*, Mimeo, Anand: Institute of Rural Management.

Bandyopadhyay, J. (1987) 'Political economy of drought and water scarcity', *Economic and Political Weekly*, 12 December.

—— (1989) *Natural Resource Management in the Mountain Environment: Experiences from the Doon Valley, India*, Kathmandu: International Centre for Integrated Mountain Development.

—— and Shiva, V. (1984) *Ecological Audit of Eucalyptus Cultivation*, New Delhi: Natraj Publishers.

Bhaskaran, S.T. (1990) 'The rise of the environmental movement in India', *Media Journal*, 37(2).

Bhatia, B. (1992) 'Lush fields and parched throats: political economy of ground water in Gujarat', *Economic and Political Weekly*, 19–26 December.

Bhatt, C.P. (1984) *Himalaya Kshetra ka Niyojan* (Renewal of the Himalaya), Gopeshwar: DGSM.
—— (1992) *The Future of Large Dam Projects in the Himalaya*, Nainital: Pahar.
Bhuskute, V.M. (1968) *Mulshi Satyagraha*, Dastane: Pune.
Brandis, D. (1884) *Progress of Forestry in India*, Edinburgh: McFarlane & Erskine.
Burra, N. (1991) *Women and Wasteland Development: A Review of NGO Experience*, mimeo, New Delhi: International Labour Office.
Caldwell, J.C. (1982) *Theory of Fertility Decline*, London and New York: Academic Press.
Calman, L. (1985) *Protest in Democratic India*, Boulder: Westview Press.
Cavalli-Sforza, L.L., Menozzi, P. and Piazza, A. (1994) *The History and Geography of Human Genes*, Princeton, NJ: Princeton University.
Centre for Science and Environment (CSE) (1985) *The State of India's Environment 1984–85: A Second Citizens' Report*, New Delhi: Centre for Science and Environment.
—— (1987) *The Wrath of Nature: The Impact of Environmental Destruction on Floods and Droughts*, New Delhi: Centre for Science and Environment.
Collins, G.F.S. (1921). *Modifications in the Forest Settlements: Kanara Coastal Tract*, Part 1, Karwar: Mahomedan Press.
Commoner, B. (1971) *The Closing Circle*, New York: Alfred Knopf.
Concerned Scholars (1986) *Bharat Aluminium Company: Gandhamardan Hills and People's Agitation*, Sambalpur: from the authors.
Conquest, R. (1968) *The Great Terror*, London: Weidenfeld & Nicolson.
Dalal, N. (1983) 'Bring back my valley', *The Times of India*, 10 July.
Dasgupta, P. (1982) *The Control of Resources*, Delhi: Oxford University Press.
Dasmann, R.F. (1988) 'Towards a biosphere consciousness', in D. Worster (ed.) *The Ends of the Earth: Perspectives on Modern Environmental History*, Cambridge: Cambridge University Press.
Dawkins, R. (1976) *The Selfish Gene*, Oxford: Oxford University Press.
Department of Science and Technology, Government of India (1990) 'Fertilizer use', in *Perspectives in Science and Technology*, Vol. 2, Science Advisory Council to the Prime Minister, Department of Science and Technology, Government of India, New Delhi: Har-Anand Publications and Vikas Publishing House.
Desai, A.R. (ed.) (1979) *Agrarian Struggles in India*, Delhi: Oxford University Press.
—— (ed.) (1986) *Agrarian Struggles in India since Independence*, Delhi: Oxford University Press.
Desmond, R. (1992) *The European Discovery of Indian Flora*, Delhi: Oxford University Press.
Devalle, S.B.C. (1992) *Discourses of Ethnicity*, Delhi: Sage Publishers.
Dhara, R. (1992) 'Health effects of the Bhopal gas leak: a review', *Epidemiologia e Prevenzione*, 52: 22–31.
Dietrich, G. (1989) 'Kanyakumari march: breakthrough despite breakup', *Economic and Political Weekly*, 20 May.
D'Monte, D. (1981) 'Time up for Tehri', *Indian Express*, 30 May.
—— (1985) *Temples or Tombs? Industry versus Environment: Three Controversies*, New Delhi: Centre for Science and Environment.
Dogra, B. (1992) *Chilika Lake Controversy: Dollars versus Livelihood*, New Delhi: from the author.
—— , Nautiyal, N. and Prasun, K. (1983) *Victims of Ecological Ruin*, Dehra Dun: from the authors.
Durning, A. (1992) *How Much Is Enough?*, The Worldwatch Environmental Alert Series, New York and London: W.W. Norton.

Economic and Political Weekly (1991) 'Gujarat: an advertiser's supplement', 5–12 January.
Ehrlich, P. (1969) *The Population Bomb*, New York: Ballantine Books.
Elwin, V. (1964) *The Tribal World of Verrier Elwin: An Autobiography*, Bombay: Oxford University Press.
Fernandes, W. and Kulkarni, S. (eds) (1983) *Towards a New Forest Policy*, New Delhi: Indian Social Institute.
—— and Ganguly-Thukral, E. (1988) *Development and Rehabilitation*, New Delhi: Indian Social Institute.
Food and Agricultural Organization, United Nations (1984) *Intensive Multiple-Use Forest Management in Kerala*, FAO Forestry Paper 53, FAO, Rome.
Gadgil, M. (1979) 'Hills, dams and forests: some field observations from Karnataka Western Ghats', *Proceedings of the Indian Academy of Sciences*, 2(3): 291–303.
—— (1989) 'Deforestation: problems and prospects', *Wastelands News*, Supplement to SPWD Newsletter, 4(4), May–July.
—— (1993) 'Biodiversity and India's degraded lands', *Ambio*, 22 (2–3): 167–72.
—— and Guha, R. (1992) *This Fissured Land: An Ecological History of India*, Delhi: Oxford University Press, and Berkeley, Calif.: University of California Press.
—— and Iyer, P. (1989) 'On the diversification of common property resource use by the Indian society', in F. Berkes (ed.) *Common Property Resources: Ecology and Community Based Sustainable Development*, London: Belhaven Press.
—— and Malhotra, K.C. (1982) 'Ecology of a pastoral caste: Gavli Dhangars of peninsular India', *Human Ecology*, 10: 107–43.
—— and Prasad, S.N. (1978) 'Vanishing bamboo stocks', *Commerce*, 136(3497): 1000–4.
—— and Rao, P.R.S. (1994) 'A system of positive incentives to conserve biodiversity', *Economic and Political Weekly*, August 6.
—— and Subash Chandran, M.D. (1988) 'On the history of Uttara Kannada forests', in J. Dargavel, K. Dixon and N. Semple (eds) *Changing Tropical Forests*, Canberra: Australian National University.
—— and Subash Chandran, M.D. (1992) 'Sacred groves', in G. Sen (ed.) *Indigenous Vision: People of India. Attitudes to the Environment*, Delhi: Sage Publications and Delhi: India International Centre, New Delhi.
—— and Vartak, V.D. (1975) 'Sacred groves of India: a plea for continued conservation', *Journal of Bombay Natural History Society*, 72: 314–20.
—— , Pillai, J. and Sinha, M. (1989) 'Report of a study conducted on behalf of fuelwood and fodder study group', Planning Commission, Government of India.
—— ,Subash Chandran, M.D., Hegde, K.M., Hegde, N.S., Naik, P.V. and Bhat, P.K. (1990) *Report on Management of Ecosystem to the Development of Karnataka's Coastal Region*, New Delhi: The Times Research Foundation.
Galeano, E. (1989) 'The other wall', *New Internationalist*, November.
Ganguly-Thukral, E. (ed.) (1992) *Big Dams, Displaced People*, New Delhi: Sage Publishers.
Ghosh, A. (1991) 'Probing the Jharkhand question', *Economic and Political Weekly*, 4 May.
Grover Smith (ed.) (1969) *Letters of Aldous Huxley*, London: Chatto & Windus.
Guha, R. (1983) 'Forestry in British and post-British India: a historical analysis', *Economic and Political Weekly*, 29 October and 5–12 November.
—— (1989a) *The Unquiet Woods: Ecological Change and Peasant Resistance in the Himalaya*, Delhi: Oxford University Press, Berkeley: University of California Press.
—— (1989b) 'Radical American environmentalism and wilderness preservation: a Third World critique', *Environmental Ethics*, 11(1).

Guha, S. (ed.) (1992) *Agricultural Productivity in British India*, Delhi: Oxford University Press.

Hart, H.C. (1956) *New India's Rivers*, Bombay: Orient Longman.

Hartman, B. (1987) *Reproductive Rights and Wrongs*, New York: Harper & Row.

Hays, S.P. (1957) *Conservation as the Gospel of Efficiency: The Progressive Conservation Movement, 1880–1920*, Cambridge, Mass.: Harvard University Press.

—— (1987) *Beauty, Health and Permanence: Environmental Politics in the United States, 1955–85*, New York: Cambridge University Press.

Herring, R.J. (1983) *Land to the Tiller*, New Haven: Yale University Press.

Hiremath, S.R. (1987) 'How to fight a corporate giant', in A. Agarwal, D. D'Monte and U. Samarth (eds) *The Fight for Survival*, New Delhi: Centre for Science and Environment.

—— (1988) 'Western Ghats march was an education', *Deccan Herald*, 14 February.

Howard, A. (1940) *An Agricultural Testament*, Oxford: Oxford University Press.

Illich, I. (1978) *Towards a History of Needs*, Berkeley: Heyday Books.

Institute of Rural Management (1992) *Farm Forestry: A Review of Issues and Pragmatic Approaches*, mimeo, Anand: Institute of Rural Management.

Institute of Social Studies Trust (1991) *Mining in the Himalayas: Report on a Field Study in the Almora and Pithoragarh Districts*, New Delhi: Institute of Social Studies Trust.

Jain, L.C. (1983) *Textile Policy Set to Annihilate Employment in the Woollen Cottage Sector*, Delhi: Industrial Development Services.

Jain, S.C. (1989) *Paper Industry: Raw Material Scenario*, mimeo, New Delhi: Straw Products Ltd.

Jan Vikas Andolan (1990) *Jan Vikas Andolan: A Working Perspective*, mimeo.

Jeffrey, R. (1992) *Politics, Women and Well Being: How Kerala Became a 'Model'*, London: Macmillan.

Jodha, N.S. (1986) 'Common property resources and rural poor in dry regions of India', *Economic and Political Weekly*, 5 July.

—— (1990) *Rural Common Property Resources: Contributions and Crisis*, Foundation Day Lecture, New Delhi: Society for Promotion of Wastelands Development.

Joshi, D. (1983a) 'Magnesite udyog: vinash ki ankahi kahani' (Magnesite mining: destruction's untold story), New Delhi: *Parvatiya Times*.

—— (1983b) 'Magnesite kanan se Himalaya ko bhaari kshati' (Grievous injuries to the Himalaya caused by magnesite mining), *Parvatiya Times*, 31 November.

Joshi, P.C. (1975) *Land Reform in India*, New Delhi: Allied Publishers.

Kalpavriksh (1988) *The Narmada Valley Project: A Critique*, New Delhi: Kalpavriksh.

Kanvalli, S. (1991) *Quest for Justice*, Dharwad: Samaja Parivartana Samudaya.

Kerala Sastra Sahitya Parishat, (1984) *Science as Social Activism: Reports and Papers on the Peoples Science Movement in India*, Trivandrum: Kerala Sastra Sahitya Parishat.

Kohli, A. (1987) *The State and Poverty in India*, Cambridge: Cambridge University Press.

Kothari, A., Pande, P., Singh, S. and Variava, D. (1989) *Management of National Parks and Sanctuaries in India. A Status Report*, New Delhi: Indian Institute of Public Administration.

Kothari, R. (1984) 'The non party political process', *Economic and Political Weekly*, 2 February.

Krishnan, M. (1975) *India's Wildlife in 1959–70: An Ecological Survey of the Larger Mammals of Peninsular India*, Bombay: Bombay Natural History Society.

Kumar, K.G. (1989) 'Police brutality besieges ecology: national fishermen's march', *Economic and Political Weekly*, 27 May.

Kumarappa, J.C. (1938) *Why the Village Movement?*, Wardha: All India Village Industries Association.

—— (1946) *Economy of Permanence*, Varanasi: Sarva Seva Sangh Prakashan.

Kurien, J. (1978) 'Entry of big business into fishing', *Economic and Political Weekly*, 10 October.

—— (1993) 'Ruining the commons: overfishing and fishworkers' actions in south India', *The Ecologist*, 23(1): 5–12.

—— and Achari, T. (1990) 'Overfishing along Kerala coast: causes and consequences', *Economic and Political Weekly*, 1–8 September.

Lal, J.B. (1989) *India's Forests: Myth and Reality*, Dehra Dun: Natraj Publishers.

Ludden, D. (1985) *Peasant History in South India*, Princeton, NJ: Princeton University Press.

Maitra, S. (1992) 'Provision of basic services in the cities and towns of the national capital region: issues concerning management and finance', Ph.D. dissertation, Jawaharlal Nehru University, New Delhi.

Malhotra, K.C. and Poffenberger, M. (eds) (1989) 'Forest regeneration through community protection: the West Bengal experience', *Proceedings of the Working Group Meeting on Forest Protection Committees, Calcutta, June 21–22*, Calcutta: West Bengal Forest Department.

Malhotra, K.C., Deb, D., Dutta, M., Vasulu, T.S., Yadav, G. and Adhikari, M. (1991) *Role of Non-Timber Forest Produce in Village Economy: A Household Survey in Jamboni Range, Midnapore District, West Bengal*, mimeo, Calcutta: Institute for Biosocial Research and Development.

Martinez-Alier, J. (1990) *The Environmentalism of the Poor*, research proposal, New York: Social Science Research Council.

Misra, D.N. (1984) 'Unjust blame on foresters', *The Times of India*, 28 January.

Misra, T.P. (1946) *Halt Hirakud Dam*, Sambalpur: Anti-Hirakud Dam Committee.

Mitra, A. (1979) 'Integrated strategies for economic and demographic development', *Economic and Political Weekly*, 3 February.

Modi, L.N. (1988) 'Whose forests are they anyway?', *Blitz*, 10 December.

Mukul (1993) 'Villages of Chipko movement', *Economic and Political Weekly*, 10 March.

Mundle, S. and Rao, M.G. (1991) 'Volume and composition of government subsidies in India 1977–78 to 1987–88', *Economic and Political Weekly*, 4 May.

Naidu, N.Y. (1972) 'Tribal revolt in Parvatipuram agency', *Economic and Political Weekly*, 25 November.

Nandy, A. (1987) *Traditions, Tyrannies and Utopias*, Delhi: Oxford University Press.

—— (ed.) (1989) *Science, Hegemony and Violence: A Requiem for Modernity*, Delhi: Oxford University Press.

Narain, S. (1983) 'Brutal oppression of fisherfolk', *The Times of India*, 11 October.

Narmada (1989–90) *Narmada: A Campaign Newsletter* (various issues), distributed by the Narmada Bachao Andolan, New Delhi.

National Commission of Agriculture (1976) *Report of the National Commission of Agriculture*, Vol. 9, Delhi: Ministry of Agriculture, Government of India.

National Fisherfolk Forum (1989) *Kanyakumari March: Breakthrough despite Breakup*, Cochin: National Fisherfolk Forum.

Norton, B.G. (1987) *Why Preserve Natural Variety?*, Princeton, NJ: Princeton University Press.

Paranjpye, V. (1981) 'Dam: are we damned?' in L.T. Sharma and R. Sharma (eds) *Major Dams: A Second Look*, New Delhi: Gandhi Peace Foundation, and Sirsi: Totgars Co-operative Sale Society.

Pathak, S. (1987) *Uttarakhand mein Kuli Begar Pratha* (The forced labour system in Uttarakhand), New Delhi: Radhakrishnan Prakashan.

Pattanaik, S.K., Das, B. and Mishra, A. (1987) 'Hirakud dam project: expectations and realities', in *People and Dams*, New Delhi: Society for Participatory Research in Asia.

Peoples Union for Civil Liberties (1985) *Bastar: Ek Mutbhed ki Jaanch* (Bastar: an enquiry report), New Delhi: Peoples Union for Civil Liberties.

Peoples Union for Democratic Rights (1982) *Undeclared Civil War*, New Delhi: Peoples Union for Democratic Rights.

—— (1986) *Gandhamardhan Mines: A Report on Environment and People*, New Delhi: Peoples Union for Democratic Rights.

Phule, J. (1883) (1969) *Shetkaryacha Asud: The Whipcord of the Farmer (1882–83)*, Pune: Maharashtra Sahitya and Sanskriti Mandal.

Prasad, S.N. (1984) 'Productivity of eucalyptus plantations in Karnataka', in J.K. Sharma, C.T.S. Nair, S. Kedarnath and S. Kondas (eds) *Eucalyptus in India: Past, Present and Future*, Peechi: Forest Research Institute.

—— ,Hegde, M.S., Gadgil, M. and Hegde, K.M. (1985) 'An experiment in eco-development in Uttara Kannada district of Karnataka', *South Asian Anthropologist*, 6: 73–83.

PUDR (1986) *Simlipal Report*, New Delhi: Peoples Union for Democratic Rights.

Raghunandan, D. (1987) 'Ecology and consciousness', *Economic and Political Weekly*, 5 May.

Rai, U., Mukul, S.B. and Kumar, D. (1991) *Call of the Commons: People versus Corruption*, New Delhi: Centre for Science and Environment.

Rangarajan, M. (1992) 'Forest policy in the central provinces, 1860–1914', unpublished D.Phil. dissertation, Faculty of Modern History, Oxford: University of Oxford.

Ravindranath, N.H., Shailaja, R. and Revankar, A. (1989) 'Dissemination and evaluation of fuel-efficient and smokeless Astra stove in Karnataka', *Energy Environment Monitor*, 5: 48–60.

Reddy, A.K.N. (1982) 'An alternative pattern of Indian industrialization', in A.K. Bagchee and N. Bannerjee (eds) *Change and Choice in Indian Industry*, Calcutta: K.P. Bagchee.

Roy, A.K., Seshadri, S., Ghotge, S., Deshpande, A., and Gupta, A. (1982) *Planning the Environment*, Anuppur: Vidushak Karkhand.

Sainath, P. (1993) 'Palamau tribals in army's firing range', *The Times of India*, 12 November.

Sangwan, S. (1991) *Science, Technology and Colonialism: An Indian Experience, 1757–1857*, Delhi: Anamika Prakashan.

Saraf, S. (1989) 'An odyssey to save the Sivaliks', *The Hindu*, 26 February.

Sarkar, S. (1983) *Modern India, 1885–1947*, New Delhi: Macmillan.

Saxena, N.C. (1990) *Farm Forestry in North-West India*, Studies in Sustainable Forest Management, No. 4, New Delhi: Ford Foundation.

Sengupta, N. (ed.) (1982) *Jharkhand: Fourth World Dynamics*, New Delhi: Authors' Guild.

—— (1985) 'Irrigation: traditional versus modern', *Economic and Political Weekly*, special number, August.

Shah, S.L. (1989) *Functioning of Van Panchayats in Eight Hill Districts of Uttar Pradesh: An Analysis of Present Malaise and Lessons for Future in the Context of the Proposed Van Panchayat Niyamwali 1989*, mimeo from the author, Almora.

Shankari, U. (1991) 'Major problems with minor irrigation', *Contributions to Indian Sociology*, 25(1).

Sharma, J.K., Nair, C.T.S., Kedarnath, S. and Kondas, S. (eds) (1984) *Eucalypts in India: Past, Present and Future*, Peechi: Kerala Forest Research Institute.

Sharma, L.T. and Sharma, R. (1981) *Major Dams: A Second Look*, New Delhi: Gandhi Peace Foundation and Sirsi: Totgars Co-operative Sale Society Ltd.

Shiva, V. (1988) *Staying Alive: Women, Ecology and Survival in India*. New Delhi: Kali for Women, and London: Zed Press.

—— , Anderson, P., Schucking, H., Gray, A., Lohmann, L. and Cooper, D. (1991a) *Biodiversity: Social and Ecological Perspectives*, Penang, Malaysia: World Rainforest Movement.

—— , Bandyopadhyay, J., Hegde, P., Krishnamurthy, B.V., Kurien, J., Narendranath, G., Ramprasad, V. and Reddy, S.T.S. (1991b) *Ecology and the Politics of Survival*, Tokyo: United Nations University Press, and New Delhi: Sage Publications.

Simon, J. (1981) *The Ultimate Resource*, Princeton, NJ: Princeton University Press.

Simon, J.L. and Kahn, H. (ed.) (1984) *The Resourceful Earth: A Response to Global 2000*, Oxford: Basil Blackwell.

Singh, K.S. (1992) *People of India: An Introduction*, Calcutta: Anthropological Survey of India.

Singh, N.J. (1992) 'Salt affected soils in India', in T.N. Khoshoo and B.L. Deekshatulu (eds) *Land and Soils*, New Delhi: Har-Anand Publications.

Singh, S. (1994) 'Environment, class and state in India: a perspective on sustainable irrigation', unpublished Ph.D. dissertation, Department of Political Science, Delhi University.

Somanathan, E. (1991) 'Deforestation, property rights and incentives in central Himalaya', *Economic and Political Weekly*, 29 June.

Somashekara Reddy, S.T. (1988) 'Tank irrigation in Karnataka', *Swayam Gramabhydaya*, 6(4): 1–5.

Space Applications Centre (1993) 'Environmental appraisal for sustainable eco-development through remote sensing data: a case study for a few watersheds of Chamoli district, central Himalaya', Scientific Note, Ahmedabad: Space Applications Centre.

State Watershed Development Cell (1989) *Watershed Development Programme for Rainfed Agriculture*, Bangalore: State Watershed Development Cell, Government of Karnataka.

Steward, W. (1988) *Common Property Resource Management: Status and Role in India*, mimeo, New Delhi: The World Bank.

Subash Chandran, M.D. and Gadgil, M. (1993) '"Kans" – safety forests of Uttara Kannada', in H. Brandl (ed.) *Proceedings of IUFRO Forest History Group Meeting on Peasant Forestry*, Freiburg, Germany, No. 40.

Subbarayappa, B.V. (1992) *In Pursuit of Excellence: A History of the Indian Institute of Science*, New Delhi: Tata McGraw-Hill.

Sukumar, R. (1989) *The Asian Elephant: Ecology and Management*, Cambridge: Cambridge University Press.

—— (1994) *Elephant Days and Nights: Ten Years with the Indian Elephant*, New Delhi: Oxford University Press.

Thakurdas, P. *et al.* (1944) *Memorandum Outlining a Plan of Economic Development for India*, London: Penguin.

Thurow, L. (1980) *The Zero Sum Society: Distribution and the Possibilities for Economic Change*, New York: Basic Books.

Valdiya, K.S. (1992) 'Must we have high dams in the geodynamically active Himalayan domain?', *Current Science*, 63(6): 289–96.

—— (1993) 'Uplift and geomorphic rejuvenation of the Himalaya in the Quaternary period', *Current Science*, 64(11, 12): 873–84.

Vijayan, V.S. (1987) *Keoldeo National Park Ecology Study*, Bombay: Bombay Natural History Society.

Vijaypurkar, M. (1988) 'Lessons from a march', *Frontline*, 20 February–4 March.
Vinayak, A. (1990) 'Tribals try to save Eastern Ghats', *Free Press Journal*, 19 December.
Viswanathan, S. (1984) *Organizing for Science: The Making of an Industrial Research Laboratory*, Oxford: Oxford University Press.
Voelcker, J.A. (1893) *Report on the Improvement of Indian Agriculture*, Calcutta: Government Press.
Vohra, B.B. (1973) 'A charter for the land', *Economic and Political Weekly*, 31 March.
—— (1980) *A Land and Water Policy for India* (Sardar Patel Memorial Lectures), New Delhi: Publications Division.
—— (1982) 'Proper land management', *Indian Express*, 27 December.
Von Oppen, M. and Subba Rao, K.V. (1980) 'Tank irrigation in semi-arid tropical India', mimeo, Patancheru: International Crop Research Institute for Semi-Arid Tropics.
West Bengal Forest Department (1988) *Project Report on Resuscitation of Sal Forests of South-West Bengal through People's Participation*, Calcutta: West Bengal Forest Department.
Whitcombe, E. (1971) *Agrarian Conditions in Northern India*, Vol. 1: *The United Provinces under British Rule, 1860–1900*, Berkeley, CA: University of California Press.
—— (1982) 'Irrigation', in D.Kumar (ed.) *The Cambridge Economic History of India*, Vol. 2, Cambridge: Cambridge University Press.
—— (1993) 'The costs of irrigation in British India: waterlogging, salinity, malaria' in D. Arnold and R. Guha (eds) *Nature, Culture, Imperialism: Essays on the Environmental History of South Asia*, Delhi: Oxford University Press.
Whyte, R.O. (1968) *Land, Livestock and Human Nutrition in India*, New York: F. Praeger.
Zachariah, M. and Sooryamurthy, R. (1994) *Science for Social Revolution*, New Delhi: Sage Publications.

INDEX

aankhbandi allowance 55–6
absentee landlords: growth of 26
accountability: lack of 50, 116, 138, 158; necessity for 124, 132
afforestation 116, 139, 193; local 104–5
agrarian reform: West Bengal 42
agrarian society, pre-colonial 129
agricultural interests: opposed to state forestry 153–4
agricultural productivity 25; desirable practices destroyed 142; enhancement of 27, 28, 65–6; exaggerated estimates of 51; many bypassed by improvements in 66; raised by green revolution strategy 26; stagnant 10; and water 20
agricultural science: need for change 144
agriculture: cereals and legumes 27–8; and external inputs 26–7, 28; farmland, the ecosystem perspective *18–19*, 25–8; health of farmlands ignored 17; intensification of 13; intensive, jobs in 182; need for cash income 165; state objectives post-independence 26; unthinking modern strategy 143; *see also* irrigation
Alakananda valley: afforestation work 103, 139; Sarvodaya workers, tree-felling rights 23; Tangsa village, use of hill streams 56–7
Ali, Salim 188
Ambedkar, Dr B.R. 115; opposed devolution of power 190
Amte, Baba 62, 74, 100
Andaman and Nicobar Islands: biodiversity in *150*, *156*, *157*; elite in *15*; forest-based industry *152*; profiting from mechanized fisheries *82*
Anthropological Survey of India (ASI) 133; People of India Project 179–80
appropriate technologists 99, 108–10, *111*
'Aravalli Chetna Yatra' 102
artisanal castes 58
artisans: forest-dependent 87–8; lost subsistence base during colonial rule 64; wood-workers, no raw materials for 168
Astraole fuel-efficient stove programme 52; evaluation of 52, *53*, *54*

Bahuguna, Sunderlal 74, 100, 103, 108, 109
Baker, Laurie 50
Balipal missile range, Orissa 76
Baliraja dam, Maharashtra 126, 143–4, 189; challenging state monopoly 58
ballots 124
bamboo: available to industry at throwaway prices 46, 140; factors involved in decline of 140–1; forests re-established 105; trends in use of 45–7
bamboo exhaustion 47
Bapat, Senapati 69
basket weavers 23, 47, 138–9, 168
bauxite extraction 89
Bedthi dam project (Karnataka): opposed by spice garden farmers 72–3; revived, but alternative scheme offered 187–8; successful opposition to 185–6
Behn, Mira 188
betelnut gardners: attempting integrated land development 186–7; offered alternative to revived Bedthi scheme

187–8; opposed Bedthi project 72–3, 186
Betla National Park 76
Bhakra-Nangal dam 71
Bharat Aluminium Company (BALCO) 89
Bhatt, Chandi Prasad 103, 104, 189
Bhave, Vinoba 119
Bhimasankar Sanctuary, Pune 94
Bhopal disaster 81
Bhopalpatnam-Inchampalli project 72
Bihar: dams, opposition to 72; free the Ganga movement 83; Jharkhand agitation 95; Palamau test firing range 76
biocentrism 151
biodiversity 155–6; Andaman and Nicobar Islands *150*, *156*, *157*; conservation of a challenge 156–62; decentralized system of conservation 161–2; enhancing long-term persistence of 159; locality-specific knowledge of 161; lost through commercial forestry 148; maintained by ecosystem people 141–2, 185; preservation of 91; role of cultivation and habitation in preservation of 160; Western Ghats 101, *150*, 185
biodiversity fund 162
biomass 10, 23; additional, due to Baliraja dam 144; commercial users, the profit element 137–8; fodder demand 168; from forests 148; and the Indian rural population 133, 136–7; locally collected, dependence on 139; needs of villagers 167–9; requirements met from public land 168
birth control *see* family planning
Bombay: milk for 25; textile mills 10
Brahmans 11, 12, 13, 58, 69
brick-making 67
Britain: encouraged increased cultivation and irrigation 25–6; India, as supplier and market 10–11; selective percolation of new techniques 29; suppressed competition 29

Calcutta: ecological refugees 32; encroachment on salt marshes 16–17, 35
canneries 17, 83
capital: natural and human-made 117–18, 120, 121, 122
capitalism 121–3; and democracy 124
carp culture 83
carrying capacity 177
cash crops 9, 26, 65
cattle and sheep 24–5
Centre for Development Studies, Thiruvananthapuram 50
cereals 27–8
chemical industries 16
children: as economic assets 181
Chilika lake 84
Chipko movement 23, 72, 84–5, 103, 109, 189
civil construction 49–51, *see also* dams
co-operatives 186; for tree farming 167
coastal development 16–17, *18–19*
colonial legacy 9–13
commercial forestry 41, 84; colonial 85–6; incompatibility of 163; post-independence 86; undermining subsistence economies 148
commercial plant production 174
common lands: encroached on by rural population 66; taken over as reserved forests 40–1
common-heritage regime 28
communities, local: and biodiversity maintenance 160; and bottom-up research 145; expansion of resource catchments 138–9; forest management systems 24, 104–5, 169–74; gaining community rights 161–2; little control over own resources 139–40; maintaining tanks 17; management systems, community-based 10, 12, 24, 38–41, 124; more control over local resource base 183; most appropriate to look after resource base 146; prudent resource use not always favoured 139; *see also* irrigation, small-scale; tank irrigation systems; village communities
community occupations 133
compensation: inadequate, for peasants moved for dams 20–1; for Mulshi dam 69
confederation: Indian style 48
conservation: hostile to tradition 158; omnivore interest in 92–4; scientific 110, *111*, 112; setting of priorities 156–7; urban groups, double

standards of 94
conservative-liberal-socialism 191; and the population-environment nexus 178–83; a working synthesis 123–32
consumption: capitalist countries 121; environmental impact of 178
Coorg forests 138
cost inflation: by Public Works Department 50
cotton growing 26, 27
crop residues 25, 168
crop varieties 27, 28, 49
cultivable land 63–6
cultivated land 25–6, 29

dairy products 25
dams 17, 68–76; benefits of exported 20–1; 'heroic', little opposition to 71; problems of 26; roots of conflicts over 69–70; siltation of 51, 143; *see also* named dams and projects
Dasholi Gram Swarajya Mandal (DGSM) 109; concerned with maintenance of local biomass 139; turned from struggle to reconstruction 103–4
Datye, K.R. 144
debt burden 60; reneging on 116–17, 131
debt trap 131
deficit financing 60
deforestation 24; shifting burden of 164, 165; through damming 20, *21*, 72; *see also* environmental degradation
Dehradun limestone controversy 88–9
democracy: and development of scientific knowledge 134–5; better environmental record of 21; grassroots, strengthening of 124; and power 38–45; parliamentary 13, 61
demonstrations/protests/rallies 74, 105–6
development: an alternative path *111*, 124–31; an ecosystem perspective on 16–33; biodiversity conservation a low priority 158–9; co-ordination lacking 48; economic, growth of artificial at the cost of natural 4; equitable/resource-efficient halting population growth 181–2; and forest management 155; information-intensive 130; people-oriented and environmentally sensitive 127, 129; process criticised by environmental action groups 100; resources and the elite 15; respecting biomass and knowledge 136–42; state funds for 49–54; and subsidized resource outflow 158
diversity: of crop varieties 28; incentives for maintenance 162; protection of 156–62; *see also* biodiversity
Down to Earth (magazine) 100
drinking water 22, 146, 190; control of 77

earthquakes 20, 21, 72
eco-development camps 101, 103–4, 139
eco-restoration projects 104–5
ecological change 2, 184
ecological classes: conflict between 63; *see also* ecological refugees; ecosystem people; omnivores
ecological degradation *see* environmental degradation
ecological prudence 191
ecological refugees 4, *14*, *32*, 33, 34, 97, 127, 140; in city hinterlands 32; creating shantytowns 68; killed in Union Carbide disaster 81; through profligate bamboo usage 47; victims in Kaveri river conflict 78
ecological security 174, 175
ecological theory: fragmentation of species-rich habitats 159; and long-term persistence of species 159–60
ecologists, neo-Malthusian 176–7
economic enterprises, scale of 128
economic liberalization 122–3
ecosystem people 3, 22, 33, 60, 97, 131; acknowledgement of biomass needs 23–4; bearing cost of environmental degradation 121–2; denied rights to forest produce 92–4, *93*; deprived of land by dam construction 68; empowerment of 119–20, 125; forest rights diminished 85; impoverished 66; and Jan Vikas Andolan 106–7; lack access to all capital 117; losing control of land 76; losses due to protected wildlife 94; many uses for local resources 141–2; in Narmada dam controversy 61–3; need for forest products 23; poverty of 34; practical knowledge and wisdom important 136–42; protection of sacred groves 91–2; sharing power in West Bengal 42–3, 189–90; small

fishermen 81–4; smaller families needed 130; Sudras and untouchables as 11; under British rule 64–5; under the Gandhian way 118; under Marxism 120; working closely with nature 141–2; *see also* peasants
ecosystem services 142–3
education 127, 128, 145; available to only a few Indians 12; caste-based reservations 37–8, 115; lacking in larger families 180–1; role of voluntary sector 37; and training, for the modern world 180, 181–2
elephant poaching 94
elite: in the Andaman–Nicobar chain *15*; and independence 12–13; seized political power 13
endangered species 151
energy: reduced dependence on external inputs 132; use of additional sources of 133–4
energy policy: and power wastage 135–6
Environment Protection Act (1986) 112
environmental action groups 99, 100
environmental change 129–30
environmental degradation 2, 25, 28, 95, *154*; attributed to ecosystem people 159; from roads and railways *30*; Kudremukh Iron Ore project 43, *44*, 45; of land outside nature reserves 159; in Marxist countries 120–1; a moral problem 107–8; root causes of 117–18, 122–3, 131–2; and rural poverty 100; tax on 126; through mining and quarrying 88–9, *90*, 125–6; through more intensive resource use 86; and tribal poverty 101–2
environmental impact assessment 186
environmental management 125
environmental movement(s) 185; defensiveness of 117; involve victims of environmental degradation 99; main concerns 2–3; no coherent alternative policy 112, 116; no creative contributions from 5; origins of 84–5; *see also* environmentalism
environmental rehabilitation 99, 103–5
environmental(ism/ists): against farm forestry 165–7; anti-development 5; efforts to co-ordinate groups 105–7; Gandhian 164; leftist 177–8; the major strands 98–112; political expression of 99–100; of the poor 98; and the population problem 176–8
equity 132; and ecology 189–90; and environmentalism 98; imperative of 163, 167–74
ethnic conflict 95–7
eucalyptus 141; seen as less harmful 165–6; in tree farming 165
export earnings 27
extremists 96

families: larger 180–1; smaller 130, 179, 180
family planning 177; coercion and health hazards 52, 54; compulsory 181; forced sterilization 45
farm forestry 165–7; new policy package needed 166–7
farming *see* agriculture
fertilizers: synthetic 13, 25, 26
finance 51–2
fish/fisheries 22, 81–4; clash between artisans and modern trawlers 81, *82*, 83; freshwater/inland 20, 83; Kanyakumari march 102–3; mechanization of 17, 81–3; post-independence development of 16, 17; rise in price of 28
folk ecology, and conservation 91–2; knowledge of 141–2, 160, 161
Forest Conservation Act (1980) 112
forest cover 175
Forest Department 23, 41, 138, 144–5; criticized 148–9; felling of trees 170; management of competing demands 155; scientific foresters 153; unwelcome in the countryside 86–7; view of ecosystem peoples' needs 23; and village forest protection committees 171–4
forest development *18–19*, 22–4
forest management, 10, 40–1, 66, 148, 149; and biomass needs 168–9; direction of 87; ecological imperative 155–62; efficiency imperative 162–7; imperative of equity 167–74; joint 171–4; local community systems 24, 104–5, 169–74; and the pursuit of profit 151; state, unwelcome and unwanted 152–3
forest resources 95, 152; available cheaply to industry 23, 24; change in management of 23–4; depleted 12;

resource base diminishing 86
forestry: and subsistence 169–71; clear cutting 10, 163; colonial 85–6; forest conflicts 84–8; influenced by middle- and upper-class needs 154; post-independence 86–8; and principles of conservative-liberal-socialism 149; state, frustration of tribals with 86–7; subsistence 169–71; 'sustained-yield' 162–3; traditional selection felling 162–3; *see also* village forest protection committees
forestry debate 148–54; interest groups 149, 151–4
forestry science: poor academic standard of 144–5
forests: community-managed 40, 184; functions performed 155; grazing in 24, 48–9, 84; liquidated by British 10; safety, establishment of 105; as source of industrial materials 151–2; state usurption of, opposed 152; subsistence function of 167–74; village interest in 116; *see also* reserved forests
fraud: and rural development programmes 35, 37
'free the Ganga' movement 83
fuel-efficient stoves 52, *53*, *54*; and betelnut gardners 187
fuelwood demand 23, 168

Gandhi, Mahatma 13, 115
Gandhi, Mrs Indira 43
Gandhian movement (Gandhism) 38–9, 65, 118–20, 123, 124, 130, 191; agrarian localism 153; crusading Gandhians 98, 107–8, 109–10, *111*, 112; flawed by emphasis on voluntary restraint 118–19, 130
GATT 120
genetic diversity 28, 141–2; *see also* biodiversity
genetic engineering 142
goats 25, 116; pressure on Western Ghats forests 48–9; subsidies for 48, 49
government: decentralized, Indian tradition of 38–9; new attempts at decentralization 41–3; two-faced 61; *see also* politics
granite quarrying *90*, 125–6
grazing lands *18–19*, 24–5, 66
green revolution 26; environmental consequences of 66; in low rainfall areas 65–6
groundwater: extraction exceeding recharge 77; pollution of 22, 26
Gwalior Rayons: industrial polluter 80

Halakar village: rights over timber 41, 139–40
Harihar Polyfibres: industrial polluter 55, 80
Haryana 65–6
Hazare, Anna Saheb 104
health and health care 31, 127
higher education 37–8
Hirakud dam, Orissa 70
HONEYBEE 161
Huxley, Aldous: quoted 176
hydroelectric projects 10, 17, 185–6; central India, opposition to 72; Silent Valley, halted 43; Western Ghats 20; *see also* dams; named projects

illiteracy 37, 96, 135
Indian Emergency 43, 45
Indian Forest Act (1878) 40
Indian Institute of Sciences 11
industrial pollution: largely unchecked 79–81, 116
industrial societies: demographic transition in 179, 180
industrialists: and forest ownership 154
industrialization 34, 116; imitative, criticized 107; pattern of 30–1; urban, an ecosystem perspective *18–19*, 28–33; a way forward 13
industrialization drive: options for 29–30
industry: at odds with Forest Department 148–9; could be supplied by farm forestry 165, 166; discouraged 10; forest-based 151–2, 164–5; manipulation and bribery by 31, 130–1; need for trained personnel 182
inefficiency: covered by overestimation of benefits 50–1; of development projects 50–1; in resource use 27, 33, 45, 130, 131, 143, 182; of sporadic use of state funds 51–2; of state monopolies 56, 57–8
inequity: in access to resources 3; in modern India 96

information 143; sharing of 134; state monopoly of 58–60, 135
inland waters development 17, *18–19*, 20–1
intensification 13–14; creating jobs 182; destroying crop diversity 27
intercropping 27
iron triangle alliance 34–5, *36*, 41, 191; acceptance of by the masses 35
irrigation 13, 17, 24, 34; and the green revolution 65–6; justification for projects 51; large-scale works 78 (*see also* dams); loss of water 22; problems with a state monopoly 57–8; small-scale 26, 38–40, 49; and soil conditions 48

Jan Vikas Andolan (People's Development Movement) 106–7
Japan 122
Jharkhand agitation 95
job quotas 38

Kanyakumari march: organized by National Fisherfolk Forum 102–3
Karanth, Shivaram 186, 188
Karnataka *90*; Astraole programme 52, *53*, 53, *54*; Coorg forests, uncertainty over ownership 138; electric pumps causing tension 77–8; empowerment of *mandal panchayats* and *zilla parishats* 37, 42, 124, 190; fuelwood depot *40*; Halakar village 41, 139–40; Kaveri river problem 78; Kudremukh Iron Ore project 43, *44*, 45; provision of drinking water 190; shrimp culture vs. normal fishing 83; social forestry programme, Ranebennur *taluk 93*, 139; Uttara Kannada 40–1, 184–8 (*see also* Bedthi dam project); Watershed Development Board 49, 187
Kaveri river, conflict over 78
Keoladeo National Park 92–3
Kerala 26; change in ecology and economy of fisheries 81, 83; industrial pollution 80
Kerala Sastra Sahitya Parishat (KSSP) 100, 129
knowledge: fragmentation of 58–9; locality-specific 161; practical 136–42; transmission to kin groups 58–9
Koel Karo dam 72
Kothari, Rajni 100
Koyna hydroelectric project 20, 34
Kshatriyas 11, 13
Kudremukh Iron Ore project 43, *44*, 45
Kumarappa, J.C. 188
Kumaun: mining and social conflict 88–9; *panchayat* forest rules 169–70

Land Acquisition Act 38
land reform 115, 120; failure of 64–6; little progress 37; pressure for 26
land use: British India 9; decentralized planning 174–5
landholding system: colonial 64–5; pre-colonial 64
leather 25
legumes 27–8
Literacy Mission 37
livestock industry: role of neglected 24–5; state intervention in 25

Maharashtra 20; destruction of *phad* irrigation system 39; effluent discharge 80; land for land resettlement policy 73–4; Ralegaon Shindi village 104; sacred groves in 91; Tandulwadi village, building of Baliraja dam 58, 126, 143–4, 189
Maharashtra Industrial Corporation: taken to court for polluting 80
management systems, community-based 10, 24, 38–41, 124; largely destroyed 12; pre-colonial 38
Mandal controversy 96
mandal panchayats 37, 41–2, 124, 190
Mandovi Bridge collapse 50
manipulation, bribery, corruption 31, 55, 130–1
Manipur: Churchandapur district, forests re-established 105, 140
market forces: and resource use 121, 122
Marxism 120–1, 123–4, 131, 191
Marxists, ecological 98–9, 108, 109, 110, *111*, 112
mining conflicts 88–9, *90*
Minor Irrigation Department: and the tank systems 21, 39
mixed economy 14–15
monopoly(ies): being broken by local initiative 187–8; shift away from 175; state 56–60; water, inefficiency of 57–8
Mulshi *satyagraha* 38, 68–70

Narmada river dams: controversy/ protests 2, 38, 61–3, 73–6, *106*; *see also* Mulshi *satyagraha*
natural resources 11, 120; chronic shortages 1–2; conflicts over access and control 63–8; during colonial period 130; and economic liberalization 123; full public accountability for 124; inefficiency/ wasteful management of 143; inequitable access to 3; management best at local level 145–6; management restructured 132; post-independence use 130–1; profit ploughed back locally 125; proper valuation needed 125–6; prudent management 127, 140–2, 146, 147; re-building of resource base 182–3; used to finance debt burden 60
natural resource conflict; *see* social conflict
natural systems, complex 140, 145–6; need for flexible adaptive management 136–7, 146
nature conservation *see* conservation; biological diversity
nature reserves 151; exclusion of ecosystem people 92–3, 159; management for biodiversity 157–8; *see also* biological diversity; conservation
Naxalites 65, 96
Nehru, Jawaharlal 14, 115
neo-Malthusians 176–7
NIMBYism 88–9
nitrate pollution 26, 66

Official Secrets Act: under the colonial administration 12, 59; used by state apparatus 60
omnivores 4, 33, 177–8; access to energy and resources limited 132; attempting to capture water resources 68–76; benefit from abuse of natural resources 60; British 9, 11–12; created by green revolution 66; environmentalist 94, 151; expansion of resource demands 130; favoured by forest policies 22–3; and Indian democracy 45; the iron triangle alliance 34–5, *36*; and nature conservation 92–4; opposition to Narmada movement 74–5; power of 135; pre-colonial 64; process resources from vast areas 141; resource capture by 35; resource consumption by 119; rural 77; sabotaged decentralized political institutions 41–2; strong hold over capital, natural and human-made 117, 118; subversion of democracy 38; under economic liberalization 123; under the Gandhian way 118; in urban centres 31
open-access land 10, 158; conversion to reserved forest 41; no communal rights of regulation 40–1; providing biomass needs 23, 24
oral communication: and spread of environmentalism 100–1
Orient Paper Mills: pollution by 80
overexploitation: of fish stocks *82*, 83; of groundwater 77–8
overfishing 17
overgrazing 24, 25
overpopulation 176

padayatras (walking tours) 101, 103
Palamau, Bihar: proposed test firing range 76
paper industry 137; cheap resources, high prices 45–6, 46–7, 140; industrial polluter 80
Patkar, Medha 62–3, 74, 75–6
patronage 35, 59, 97
peasants 77; driven from lands 38; Himalayan, protest at pressure on water resources 72; impoverished 64; knowledge of 58; losing control of land 26; resented dam building 68–71; *see also* ecosystem people
People's Science Movements (PSMs) 108, 129, 143
pesticides 13, 26–7, 81; pollution by 66, 122
petroleum 17, 27, 60, 118
phad irrigation system 39
planned development: created islands of prosperity 34; deterioration of 116
plantations, monocultural 151, 163, 165; failure of 141
poaching 92, 94
politics: decentralized 38–9, 41–3, 189, 190, 191
pollution 20, 22; a behemoth out of control 116; of coastal waters 17, 102;

from tanneries 25; industrial 79–81, 116; polluter pays principle 126–7; problem solved by dilution 21–2; sea 43; solution to 21–2
pollution control: legislation, Indian style 79–81; and regulation 56, 127
popular participation: potential force of 169–71
Population Bomb, The, P. Ehrlich: quoted 176–7
population categories: boundary problems 4–5; *see also* ecological refugees; ecosystem people; omnivores
population growth: deceleration of 179–80, 181–2; and environmental degradation 177–8; governed by self-interest 178–9, 182; upward 129–30, 178–9
power generation 17, 20; small-scale 56–7; *see also* hydroelectric projects; named projects
print media: and environmentalism 100, 103
private enterprise: encouraged to deliver services 37, 127–8; socially responsible 132; subsidized by British 14; under state regulation 14–15
Private Trees Protection Act: Nilgiris district 55–6
Project Tiger 151
protein and protein deprivation 17, 27–8
Public Works (Waste?) Department 49–54
Punjab 65–6

rainforest: destroyed for eucalypts 50–1; resources of 141; in sacred groves 185; saved 73
Ralegaon Shindi village, Maharashtra, eco-restoration through self-help 104
Ranebennur *taluk*, Karnataka: social forestry programme *93*, 139
Rauwolfia 141
rayon industry 80, 116
refugia 159, 185
regulation/regulators: gain from profligacy 137–8; and the iron triangle 55–6
rehabilitation, environmental 103–5
research: bottom-up strategy 145; species for tree farming 167
reserved forests 10, 40–1, 66; and biomass needs 168–9
reservoirs *see* dams
resettlement: of dam evacuees 71, 72, 73–4, 117
resource use 15, 134, 162–7; inefficient 27, 33, 45, 130, 131, 143, 182; intensification of, paid for by British 13–14; and subsidies 27; more efficient in capitalism 121; profligate, costs passed on 47; refashioning of 13; responsibility of bureaucracy 55–6
resource wastage 48–9, 50–2, 54–6
resources 22, 45; consumption among British elite 9; exhaustive use of 22–3, 158; local control of 124; more equitable access to needed 126; resource base dwindling 2; spent decoupled from services delivered 47–54; subsidized for industry 30; and subsistence 28–9; switching between 137–8
rewards: for custodianship of biodiversity/knowledge 161–2
Rihand dam, Uttar Pradesh 71
rotation 27
royalties: and biodiversity 162
rural social activists 155; work among ecosystem people 152–3

Saab, Abdul Nazir 190
sacred groves 91–2; Uttara Kannada 185
sal forests: Arabari, recovery of 170–1, 172
salination 26, 48, 51
Salt Lake City, Calcutta 35
Sardar Sarovar dam 61–3; opposition to 73–6
Sarvodaya movement 23, 104; *see also* Gandhian movement
satellite imagery 1, 129; showing effects of environmental rehabilitation 103–4
Save the Nilgiris march 101
Save the Sivaliks march 101–2
Save the Western Ghats March 101
science/technology: attitudes to 109; demystifying of 129; enhancement of ecosystem services 142–3; environment-friendly 145; new ways forward 129
scientific conservation 110, *111*, 112
scientific foresters 153
scientific management: bogus claims! 143

sea pollution 43
seed banks 28
separatist movements 95–6
shantytowns 68, 81
Sharavathy waterfalls: used for hydroelectric power 185–6
shrimp culture 83, 84
Silent Valley hydroelectric project: halted 43, 73
siltation 83; of dams 51, 143; from Kudremukh Iron Ore project 43; of tank systems 39, 49
Simlipal Tiger Reserve, Orissa 92
Singh, V.P. 61
slum clearance: conflict over 68
social activism 188–9; constructive *see* Uttara Kannada; social action groups 99, 100, 107; social forestry programmes 139; *see also* rural social activists; Gandhian movement; Marxists, ecological
social conflict 2; caused by dam building 68–76; and fishing 81–4; and forestry 84–8; and mining 88–9; Narmada controversy 61–3; and nature conservation 92–4, *93*; over natural resources 63–8; and religious interests 91–2; and water control 76–81
soil conservation 143; Alakananda valley 103
soil erosion 143, 148
soils: fertile, Haryana and Punjab 65–6; tropical, nutrient-poor 10, 24, 84
Soule, Michael: on ecologists 184
Soviet Union: lack of open discussion 134–5; *see also* Marxism; state socialism
SRISHTI 161
state: as guarantor of environmental protection 110, 112; seen as source of free handouts 35, 37; usurping rights of local communities 139–40
state apparatus 12; begged for overseas resources 32–3; centralized 135; and the development process 15; and economic liberalization 122; monopolistic control 56–60; not curbed by leftist state governments 120; and preservation of biodiversity 160–1; salaries, perks and state funds 60; spending and regulation by 47; treatment of renewable resources 46; unconcerned over resource wastage 51–2; unsatisfactory performance unchecked 54–6
state forestry *see* forestry; forest management
State Pollution Board: sued 80
state socialism: failure of 116, 127
Subarnarekha dam, Bihar 72
subsidies 35, 116, 122–3, 158; and the development process 15; discontinuation of 130; for farming inputs 26–7; more equitable distribution needed 126; and omnivores 60; to bring cheap water to cities 78–9
subsistence 12; and the forests 155, 167–74; importance of forest grazing 84; and resources 28–9; *see also* ecosystem people
Sudras 11, 59
sustainable use 41, 46, 47, 100, 132, 161, 184; motivation for 137–8

Tamil Nadu: Kaveri river problem 78; Private Trees Protection Act, Nilgiris 55–6; protests against pollution from tanneries 80–1
tank irrigation systems 17, 26, 39; collapse of 21; silting of 39, 49
targetbaji 51–4
Tata industrial house: acquisition of cheap land for power projects 20; beginning to show environmental concern 189; proposed 'Integrated Shrimp Farming' project 84; provoked Mulshi protest 38, 68–9
Tata, J.N. 11
taxes: agricultural 10; environmental 126; land tax 9, 11–12; levied by water-lords 83
teak 141
technology(ies): capital-intensive 30; import of 118; and India 134; vs. pastoral-agrarian system 128–9; *see also* science/technology
Tehri dam: long-standing opposition to 72; suspension of refused 43
Thippegondanahalli reservoir: housing development scandal 67–8
timber harvesters 151–2, 155
Times of India, The: on Mulshi *satyagraha* 69–70
trawlers: effects of 81–3

tree farming 165–7; disenchantment with 166; new policy package needed 166–7
tribal regions: affected by protected species 94; dependence on forests 85; impoverishment of 95
Tungabhadra project, Karnataka 55, 71
Tungabhadra river: pollution of 55, 80

Union Carbide disaster 81
untouchables 11
Upper Krishna Reservoir: religious opposition to 91
urban industrial enclaves 10, *18–19*, 20, 28–33
urban land: scandals surrounding development of 67–8; value of 67
Uttara Kannada, Karnataka 40–1; 184–8; successful village forest councils 41; *see also* Bedthi dam project
Uttarakhand Sangarsh Vahini 109

Vaisyas 11–13
van panchayats 169–70, 173–4
village common lands 24, 63
village communities 153; activities of, pre-independence 49; excluded by industrial forestry 163; *see also* ecosystem people
Village Forest Act (1926) 41
village forest councils 41
village forest protection committees (FPCs) 171–4, 189–90
village forests 169–70, 173–4
village grazing lands 66
village handicrafts: some support for 30
village *panchayats* (councils) 126
village republics (Gandhian model) 115, 124; rejected 13
village society, pre-colonial: call for return to 107–8
villages: displaced by dams 2, 20; shifted from protected areas 92, 158
Vishnuprayag project: resistance to 72
Vohra, B.B. 110
voluntary agencies: services delivered by 37, 127–8
voluntary restraint 118–19, 130

water (resources): for dilution of pollution 21–2; misused 57–8; post-independence mobilization of 20; as resource and sink 76–81; source of much social conflict 2, 68–76
waterlogging 17, 26, 48, 51, 66
watershed development programmes (integrated) 49, 187
West Bengal 26; decentralization of power 42–3, 189; power-sharing with ecosystem people 189–90; recovery of Arabari sal forests 170–1; success of village forest protection committees 171–2, 189–90
West Coast Paper Mill: bamboo at throwaway prices 46; change to other resources 47
Western Ghats 20, 34, 141; biodiversity in 101, *150*, 185; *Dipterocarpus* species 185; goat grazing 48–9; *see also* Karnataka; Maharashtra
wild animals: damage by 94, 158
wilderness conservationists 149–51, 153, 154, 155
wilderness protection movement 110, 112, 116
Wildlife Protection Act (1972 and 1991) 112
women: benefiting from forest management 173; involved in protests 89; political power for 190
World Bank 75, 117, 120; loan to bring cheap water to cities 78–9

zamindari settlements 64
zilla parishats 37, 41–2, 124, 190